装备科技译著出版基金

无人机系统成像与感知
——部署与应用

Imaging and Sensing for Unmanned Aircraft Systems:
Deployment and Applications

［巴西］瓦尼亚·V. 埃斯特雷拉（Vania V. Estrela）
［印］裘德·赫曼（Jude Hemanth）
［日］奥萨姆·邵图米（Osamu Saotome） 编著
［瑞典］乔治·尼古拉科普洛斯（George Nikolakopoulos）
［阿联酋］罗伯托·萨巴蒂尼（Roberto Sabatini）

吴剑旗　李　晨　译
李　辉　王传声　审校

国防工业出版社

·北京·

著作权合同登记　图字:01-2023-2212号

图书在版编目(CIP)数据

无人机系统成像与感知:部署与应用/(巴西)瓦尼亚·V. 埃斯特雷拉(Vania V. Estrela)等编著;吴剑旗,李晨译. —北京:国防工业出版社,2024.4
书名原文:Imaging and Sensing for Unmanned Aircraft Systems:Deployment and Applications
ISBN 978-7-118-13166-6

Ⅰ.①无… Ⅱ.①瓦… ②吴… ③李… Ⅲ.①无人驾驶飞机-成像系统 Ⅳ.①V279

中国国家版本馆CIP数据核字(2024)第060506号

Imaging and Sensing for Unmanned Aircraft Systems:Deployment and Applications by Vania V. Estrela, Jude Hemanth, Osamu Saotome, George Nikolakopoulos and Roberto Sabatini
ISBN:978-1785616426
Original English Language Edition published by The Institution of Engineering and Technology, Copyright © The Institution of Engineering and Technology 2020, All Rights Reserved.
本书简体中文版由Institution of Engineering and Technology授权国防工业出版社独家出版发行,版权所有,侵权必究。

※

国防工业出版社出版发行

(北京市海淀区紫竹院南路23号　邮政编码100048)
雅迪云印(天津)科技有限公司印刷
新华书店经售

*

开本710×1000　1/16　插页5　印张15¼　字数260千字
2024年4月第1版第1次印刷　印数1—2000册　定价146.00元

(本书如有印装错误,我社负责调换)

国防书店:(010)88540777　　书店传真:(010)88540776
发行业务:(010)88540717　　发行传真:(010)88540762

无人机成像与感知丛书译审委员会

主　任	吴剑旗				
副主任	李　晨	李　辉	王传声		
委　员	蒋大刚	郭拓荒	杨云志	丁　群	陈赤联
	徐春叶	黄大庆	昌　敏	牛轶峰	石　亮
	赵　钰	施孟佶	陈　彦	陈华伟	汪子君
	魏明珠	方黎勇	孙　彬	黄　健	张天良
	乔　伟	马煜然	惠国保	高劲飞	杨　巍
	丁文锐	冯　慧	刘中杰	孟祥玲	苏纪娟
	吴永亮	王燕宇	曹　菡	秦　乾	党帅军
	李一航	张　鑫	张瑞雨	徐玉龙	孙仁武
	姬鹏飞	毛登森	张　明		

丛 书 序

计算机视觉也称为机器视觉,是人工智能技术的应用之一,囊括很多能够理解图片图像和视频图像的算法,是许多创新型关键技术的基础,作为一种重要的感知手段,用途极为广泛。计算机视觉与无人机系统的结合,极大地拓展了无人机系统的用途。

近年来,我国持续扩大低空空域开放,促进了通航产业发展,无人机系统作为通航产业的重要内涵,发展尤为迅猛。目前,无人机系统可装载或挂载多样化任务载荷,可执行各种复杂任务,现已在电力、安防、测绘、能源、植保等领域中得到广泛应用;无人机在战场的使用将推动有人/无人协同作战的发展,极大改变未来作战方式和形态,各种军用无人机系统也在实战中发挥了不可替代的作用。

纵览无人机系统技术的发展,成像与感知技术正逐渐成为驱动无人机系统创新发展的关键支撑技术,拥有巨大的发展前景,无人机作战技术应用趋于智能化发展,具有感知智能化、判断智能化、决策智能化、打击智能化等特点。

原著作者团队包括马绍尔群岛计算机物理系统集团的创始主席和自主智能航空航天系统实验室的主任 Roberto Sabatini、巴西航空技术研究所的教授 Osamu Saotome 等专家学者,在航空航天领域具有丰富的研究经历和成果,全书围绕无人机系统成像能力和传感器集成部署,深度解析无人机系统引领的技术革命,系统展示了自主智能体探测与感知方面的成就和面临的挑战。

本书译者团队来自中国电科集团下属研究院所、电子科技大学航空学院和南京航空航天大学无人机学院等国内无人机系统领域具有领先视野的院所和高校。内容力求深入浅出、图文并茂,全景解析无人机系统在成像与感知领域的技术成果和发展趋势。本书介绍了无人机系统的图像和传感技术基础,阐述了机器视觉和数据存储在无人机系统的现状,视觉系统在无人机态势感知和探测、智能导航和姿态估计的应用,对无人机系统控制模型和仿真、多传感器融合导航、基于视觉的结构健康监测等技术进行了详细介绍,内容涵盖单体无人机系统的结构、姿态、导航及任务载荷的技术积累和展望。本书的出版将对无人机系统领域,尤其是计算机视觉技术在无人机应用中的理论基础和应用基础起到重要的推动作用,并为无人机系统探测与感知领域的研究人员和学生提供系统全面的知识资源,将进一步推进无人机系统在军事和民用领域的飞速发展。

院士

编著者简介

瓦尼亚·V. 埃斯特雷拉(Vania V. Estrela)，弗鲁米嫩塞联邦大学(UFF)电信系的教员/研究员，UNICAMP 大学的访问学者，主要研究领域包括生物医学工程、电子仪器、建模/模拟、可持续设计、多媒体、人工智能、遥感、STEM 教育、环境和数字包容。曾担任电气和电子工程师协会(IEEE)、爱思唯尔、美国计算机学会、IET、斯普林格-弗拉格和 MDPI 的评审专家。她获得理学硕士和博士学位，拥有丰富的项目经理、博导经验，以及做为书籍和特刊的编辑。

裘德·赫曼(Jude Hemanth)，印度卡伦亚大学欧洲经济学院的副教授，美国电气和电子工程师协会深度学习工作组成员，国际一些参考期刊的副主编和编辑委员会成员。

奥萨姆·邵图米(Osamu Saotome)，巴西国际航空技术学院(ITA)的教授。曾参与巴西空军、INPE、IEAv(法国、瑞典、美国、加拿大和日本)的几个国际研究和合作项目。

乔治·尼古拉科普洛斯(George Nikolakopoulos)，瑞典鲁里亚理工大学(LTU)计算机科学、电气和空间工程系机器人和自动化教授。欧洲委员会非洲实时环境监测信息系统科学委员会的成员。在管理由欧盟、欧洲航天局、瑞典和希腊国家研究部资助的欧洲和国家研发和投资(R&D&I)项目方面拥有丰富的经验。

罗伯托·萨巴蒂尼(Roberto Sabatini)，皇家墨尔本理工大学(RMIT)工程学院航空航天工程和航空学教授，RMIT 计算机物理系统集团的创始主席和自主智能航空航天系统实验室(劳伦斯·瓦克特中心)主任。在航空航天、国防和运输领域拥有超过 25 年的经验，包括先进的学术和军事教育、广泛的研究和飞行试验实践，以及在欧洲、美国和澳大利亚的大学和研发组织中先后担任技术和运营领导职务。除了拥有航空航天/航空电子系统(克兰菲尔德大学)和卫星导航/地理空间系统的博士学位(诺丁汉大学)，萨巴蒂尼持有试飞工程师(快速喷气式)、私人飞行员(固定翼飞机)、远程飞行员(多旋翼无人飞机)执照。在其整个职业生涯中，他成功地领导了许多关于航空航天、国防和运输系统的工业和政府资助的研究项目，并撰写或共同撰写了 250 多份经同行评审的国际出版物和 100 多份研究/飞行试验报告。

萨巴蒂尼是特许专业工程师、工程主管和澳大利亚工程师学会会员。此外，他还是皇家航空学会会员、皇家航海学会会员(FRIN)、电气和电子工程师协会(IEEE)高级会员、美国航空航天学会(AIAA)高级会员和武装部队通信和电子协会(AFCEA)终身会员。他获得了各种科学和专业奖项,包括 ADIA 年度科学家奖(2019 年)、北约研究和技术组织科学成就奖(2008 年)、SAE Arch T. Colwell 奖(2015)、SARES 科学奖(2016)和诺斯罗普·格鲁曼职业奖学金(2017)。他是《航空航天科学进展》的航空电子编辑、美国电气和电子工程师协会(IEEE)汇刊《航空航天和电子系统学报》的技术编辑、《智能和机器人系统杂志》的高级编辑、《航空航天科学和技术》的助理编辑以及《导航杂志》的助理编辑。此外,他还担任美国电气和电子工程师协会(IEEE)航空电子系统小组副主席、美国国家航空航天局(NASA)无人机系统(UAS)交通管理(UTM)合作测试计划成员,以及国际民用航空组织(ICAO)航空环境保护委员会(CAEP)、影响和科学小组(ISG)的澳大利亚国家代表。

萨巴蒂尼从事航天、运输和国防应用的智能自动化和自治系统的研究,其研究领域包括航空电子设备和空间系统;通信、导航和监视/空中交通管理;制导、导航和控制(GNC);全球导航卫星系统;无人机系统(UAS)和 UAS 交通管理(UTM);国防 C^4ISR 和电子战系统;人机系统和可信自主。其研究导致了重大发现,包括创新的导航和制导技术;最佳控制和轨迹优化:全球导航卫星系统完整性增强;激光/光电传感器;实验飞行试验技术和仪器:UAS 感知和回避;认知人机系统(自适应人机界面和交互,以实现可信的自主性和增强的人类性能)。

第2卷 前言

无人驾驶飞行器(UAV)简称无人机,也称无人机系统(UAS)或遥控飞机系统(RPAS),是一种没有飞行员的飞机。它的飞行可以由车辆中的计算机自主控制,也可以通过远程控制。它们可以帮助执行大量的任务,如在偏远地区的监视、救灾、医疗等。无人机具有独特的穿透区域的能力,这对于有人驾驶飞行器来说可能太危险了。使无人机自主化需要解决不同学科的问题,如机械设计、航空、控制、计算机科学、传感器技术和人工智能等。

无人机网络-物理系统(CPS)包括:① 子系统和接口;② 机电一体化框架(航空);③ 地面控制站[1-2]。为了达到期望的自主性水平,航空电子设备具有优化设计、多维传感器框架和启动。一架完全自主可控的无人机可以:① 获取有关地球的数据;② 在没有人为干扰的情况下为全方位的时间框架工作;③ 在没有人为协助的情况下飞越其工作区域;④ 避免个人及其财产面临的危险情况。

无人机需要定制作战系统(CPS),这些无人机定制作战系统无疑是多维度的,特别是考虑到它们的所有潜力、广泛和多个作战规模以及社会影响。

无线传感器网络(WSN)和视觉传感器与执行器网络(VSAN)是 CPS 中无人机的重要组成部分。UAV-CPS 对 WSN 的设计提出了许多限制,如可扩展性、容错性、操作环境、硬件限制、生产预算、网络拓扑和功耗[3-18]。这些挑战需要进行研究,以解决传感器在信息获取和处理中可能存在的协作问题。

因此,传感器和执行器节点的基本要求是在有限的能源下运行。网络运营应该成功且积极有效,从而有足够的时间来执行必要的应用程序。

无线多媒体传感器网络(WMSN)是由无线连接的传感器节点组成的网络,具有多媒体组件,可处理诸如侦察密集传感器网络、电子政务控制系统、尖端医疗保健提供、老年人援助、远程医疗和灾害补救及控制。在这些应用中,多媒体支持有可能提高收集的证据水平、扩大覆盖范围并允许多分辨率视图(与标量数据的测量相比)。WMSN 除了无线传感器网络的特点和挑战之外,还具有实时多媒体信息的性质,如高带宽需求、实时提供、可接收端到端延迟、合适的抖动和丢帧率。同样,WMSN 有无数不同的限制资源,包括功率、带宽、处理能力、数据速率、内存和缓冲器的大小,这是由于传感器的尺寸非常小和通常海量数据的应用程序的多媒体性质造成的。

因此,为了满足服务质量(QoS)的需要并有效地管理网络的硬件和软件资源、WMSN 特性以及覆盖和安全等其他问题（图 P.1）。这些特性肯定应该在不同的通信协议层上考虑，而这些问题将在后续章节中详细介绍和讨论。此外，考虑到 VSAN 数据中相对较高的冗余，WMSN 有额外的要求，如节点内多媒体处理（如分布式多媒体编码和数据压缩）、专用程序的 QoS 需求和多媒体网络内计算模块（如多种数据融合的可能性、存储管理、板外处理和聚合）。

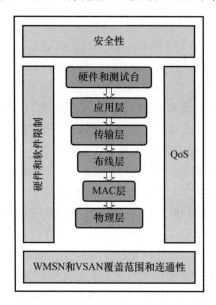

图 P.1 从网络角度看的无人机信息 - 物理系统

就频段而言，物理层可以使用三个技术组：窄带、扩频和超宽带（UWB）。这些技术依赖于多种调制结构和带宽考虑，具有不同的标准协议（IEEE802.15.1 蓝牙、IEEE802.15.4ZigBee、IEEE802.11Wi-Fi、802.15.3aUWB）。ZigBee 是 WS 中最常用的标准无线协议，因为它具有轻量以及低成本和低功耗特性，在 2.4GHz 下可实现数据速率高达 250kbit/s、编码效率高达 76.52% 和协议范围达到 10~100m。尽管如此，ZigBee 不适合高速率数据使用，如 WMSN 上的多媒体流以及确保应用程序 QoS。相反，蓝牙和 Wi-Fi 等其他标准具有更高的数据速率和代码效率，但它们需要额外的功率。

本卷分为以下几章：

第 1 章：作为新技术测试台和工业 5.0 基础的无人机信息 - 物理系统

第 1 章主要介绍无人机 - CPS 的几个问题和目前的研究现状。讨论了几个趋势和需要，以促进读者对下一章进行评论，并值得深思。

第 2 章：无人机系统人为因素和人机界面设计

无人机 CPS 的显著优点是，它们能够在完成例行工作的同时，承担人类的职责和目标。同时，他们也独立地做出一些决定和执行一些任务，因此，人和机器必须协作。尽管这些功能提供了显著的回报,但要完全掌握辅助人机交互的合适方法,仍然需要付出巨大的努力。

第3章:无人机中的开源软件和开源硬件

将低成本、开源组件与多个传感器及执行器一起使用,在工作量和成本方面是一个相当大的挑战。因此,在无人机 CPS 中采用 OSS 和 OSH 是一种理想的解决方案。在有可能的情况下,应在设计中使用可用的 OSH 和 OSS,而不受硬件框架类型和操作系统类型的限制。

第4章:基于多重描述编码(MDC)的无人机 MIMO UWB – OSTBC 系统瑞利信道图像传输

正交频分复用(OFDM)UAV – CPS 的性能可以通过添加信道编码(纠错码)来提高,以识别和纠正在数据传输过程中发生的错误。与传统的无线信道模型相比,UWB 信道的巨大带宽可以在多个 MAV 应用中产生新的影响。当需要高精度时,基于 IEEE 802.15.4a 的 UWB 技术具有多种用途,以实现稳定和导航,这绝对是无人机室内定位的挑战。

第5章:为摄影测量、遥感和计算机视觉记录飞行状态的低空无人机飞行图像数据库

在条件多变的非结构化环境中飞行无人机具有挑战性。为了支持更好算法的开发,提出了一个多用途数据集,用于给定巴西环境下的低空无人机飞行,作为和其他航空电子任务的定位基准,以便根据稳健性和泛化性评估计算机视觉程序有无地标的深度评估基线。这一发展阶段有助于推进未来与遥感(RS)模块的集成,从而为分析带来更多光谱信息。

第6章:通信要求、视频传输、通信链路和联网无人机

UAV – CPS 涉及大量关于网络的知识——具体来说,是关于飞行 Ad – hoc 网络(FANET)。通过 UAV – CPS 的高维多媒体数据流的流量往往呈指数级增长,这一事实为未来的研究方向提出了几个问题。

第7章:无人机多光谱与高光谱成像:现状与展望

纹理是识别图像中的目标或感兴趣区域(ROI)的重要特征,如它通常从卫星图像中进行图像的分类。无人机图像利用超高空间分辨率,这表明纹理也是最重要的信息来源。然而,无人机图像中的纹理很少用于监视。此外,合并地面高光谱数据可以弥补无人机传感器的有限波段,提高分析的估计精度。因此,本章的目标是:① 探索基于无人机的多光谱图像;②通过高光谱信息提高几种类型的评估精度。

第8章:无人机航拍成像和基础设施的重构

本章将介绍配备摄像头的无人机在基础设施空中目检中的应用,如施工项目、建筑物、桥梁、森林保留地和其他需要频繁进行空中目检的基础设施。这些

UAV-CPS可以经常检查各种站点,监控正在进行的工作并生成这些环境的详细三维重建以供进一步利用。

第9章:深度学习无人机系统中超分辨率成像的替代方案

无论星载传感器、云端、遥感、计算智能化和通信技术的发展如何复杂,超分辨率(SR)的需求仍将持续相当长一段时间。例如,复杂的医疗保健、天文学和救灾领域,获取图像是昂贵的,这种情况将持续发生。

第10章:无人机系统的体验质量和服务质量

QoS和QoE(除其他定性性能指标外)将发挥关键作用,推动UAV-CPS各个阶段的进一步改进。

本书旨在为当前和未来的无人机CPS应用提供参考。它将展示无人机成像能力和传感器集成的基本方面、正在进行的研究工作、所面临的困难和挑战。

Vania V. Estrela

Jude Hemanth

Osamu Saotome

George Nikolakopoulos

Roberto Sabatini

参考文献

[1] Estrela, V. V., Saotome, O., Loschi, H. J., et al. 2018. Emergency response cyber-physical framework for landslide avoidance with sustainable electronics. Technologies. 6:42. doi:10.3390/technologies6020042

[2] Estrela, V. V., Monteiro, A. C. B., França, R. P., Iano, Y., Khelassi, A., and Razmjooy, N. 2019. Health 4.0: applications, management, technologies and review. Med. Technol. J. 2(4):262-276. doi:10.26415/2572-004xvol2iss1p262-276

[3] Ezequiel, C. A. F., Cua, M., Libatique, N. C., et al. UAV aerial imaging applications for post-disaster assessment, environmental management and infrastructure development. In Proceedings of the 2014 IEEE International Conference on Unmanned Aircraft Systems (ICUAS), Orlando, FL, USA, 2014; pp. 274-83.

[4] Lewkowicz D. J., and Ghazanfar A. A. 2009. The emergence of multisensory systems through perceptual narrowing. Trends Cogn. Sci. 13(11):470-478. doi:10.1016/j.tics.2009.08.004

[5] Zmigrod, S., and Hommel, B. 2010. Temporal dynamics of unimodal and multimodal feature binding. Atten. Percept. Psychophys. 72(1):142-152. doi:10.3758/APP.72.1.142

[6] Nitti, D. O., Bovenga, F., Chiaradia, M. T., Greco, M., and Pinelli, G. 2015. Feasibility of using synthetic aperture radar to aid UAV navigation. Sensors. 15:18334-18359.

[7] Park, C., Cho, N., Lee, K., and Kim, Y. 2015. Formation flight of multiple UAVs via onboard sensor information sharing. Sensors. 15:17397 – 17419.

[8] Blaauw, F. J., Schenk, H. M., Jeronimus, B. F., *et al.* 2016. Let's get Physiqualan intuitive and generic method to combine sensor technology with ecological momentary assessments. J. Biomed. Inform. 63:141 – 149.

[9] Kang, W., Yu, S., Ko, S., and Paik, J. 2015. Multisensor super resolution using directionally – adaptive regularization for UAV images. Sensors. 15:12053 – 12079.

[10] Karpenko, S., Konovalenko, I., Miller, A., Miller, B., and Nikolaev, D. 2015. UAV control on the basis of 3D landmark bearing – only observations. Sensors. 15:29802 – 29820.

[11] Yoon, I., Jeong, S., Jeong, J., Seo, D., and Paik, J. 2015. Wavelengthadaptive dehazing using histogram merging – based classification for UAV images. Sensors. 15,6633 – 6651.

[12] Wen, M., and Kang, S. 2014. Augmented reality and unmanned aerial vehicle assist in construction management. Comput. Civil. Building. Eng. 1570 – 1577. doi: 10.1061/9780784413616.195

[13] Li, H., Zhang, A., and Hu, S. A 2015. Multispectral image creating method for a new airborne four – camera system with different bandpass filters. Sensors. 15:17453 – 17469.

[14] Aroma, R. J., and Raimond, K. 2017. A novel two – tier paradigm for labeling water bodies in supervised satellite image classification. Proc. IEEE 2017 International Conference on Signal Processing and Communication(ICSPC), Coimbatore, India, 384 – 388.

[15] Chiang, K. – W., Tsai, M. – L., Naser, E. – S., Habib, A., and Chu, C. H. 2015. New calibration method using low cost MEM IMUs to verify the performance of UAV – borne MMS payloads. Sensors. 15:6560 – 6585.

[16] Roldan, J. J., Joossen, G., Sanz, D., del Cerro, J., and Barrientos, A. 2015. Mini – UAV based sensory system for measuring environmental variables in greenhouses. Sensors. 15:3334 – 3350.

[17] Gonzalez, L. F., Montes, G. A., Puig, E., Johnson, S., Mengersen, K., and Gaston, K. J. 2016. Unmanned aerial vehicles (UAVs) and artificial intelligence revolutionizing wildlife monitoring and conservation. Sensors. 16:97.

[18] Razmjooy N., Ramezani M., and Estrela V. V. (2019) A solution for Dubins path problem with uncertainties using world cup optimization and Chebyshev polynomials. In: Iano Y., Arthur R., Saotome O., Vieira Estrela V., and Loschi H. (eds) Proc. BTSym 2018. Smart innovation, systems and technologies, vol 140. Springer, Cham. doi: 10.1007/978 – 3 – 030 – 16053 – 1_5

目 录

第1章 作为新技术测试台和工业5.0基础的无人机信息-物理系统 001
  1.1 引言 001
  1.2 云计算 004
  1.3 集体无人机学习 006
  1.4 人类计算、众包和呼叫中心 007
  1.5 开源和开放访问的资源 008
  1.6 挑战和未来的方向 009
  1.7 小结 012
  参考文献 012

第2章 无人机系统人为因素和人机界面设计 022
  2.1 引言 022
  2.2 无人机系统人机界面功能 025
    2.2.1 可重新配置显示器 026
    2.2.2 感知与规避 026
    2.2.3 任务规划和管理 026
    2.2.4 多平台协作 026
  2.3 地面控制站人机界面元素 028
  2.4 人因计划 030
    2.4.1 需求定义、获取和改进 033
    2.4.2 任务分析 034
    2.4.3 分层任务分析 034
    2.4.4 认知任务分析 035
    2.4.5 关键任务分析 035
    2.4.6 操作顺序图 036
    2.4.7 系统设计和开发 036
    2.4.8 设计评估 037

  2.4.9 验证与确认 ··· 039
 2.5 未来的工作 ··· 040
 2.6 小结 ·· 040
 参考文献 ·· 041

第3章 无人机中的开源软件和开源硬件 ·· 043

 3.1 引言 ·· 043
 3.2 开源软件 ·· 044
 3.3 开源 UAS ··· 045
 3.4 通用消息协议 ·· 048
 3.5 GCS 软件 ··· 050
 3.6 处理软件 ·· 051
 3.7 操控员的信息与通信 ·· 052
 3.8 开源平台 ·· 054
 3.9 未来工作 ·· 054
  3.9.1 OSH 挑战 ·· 055
  3.9.2 开放数据 ·· 055
  3.9.3 云数据中心 ·· 055
  3.9.4 UAV-CPS 中的众包数据 ·· 056
  3.9.5 无人机群的控制 ·· 056
 3.10 小结 ·· 057
 参考文献 ·· 057

第4章 基于多重描述编码(MDC)的无人机 MIMO UWB-OSTBC 系统瑞利信道图像传输 ··· 060

 4.1 引言 ·· 060
  4.1.1 平坦瑞利衰落信道的效率 ······································ 064
 4.2 多重描述编码(MDC) ·· 064
 4.3 多输入多输出 ·· 066
 4.4 分集 ·· 068
 4.5 仿真结果 ·· 068
 4.6 讨论和未来发展趋势 ·· 073
 4.7 小结 ·· 075
 参考文献 ·· 076

第5章 为摄影测量、遥感和计算机视觉记录而生的低空无人机飞行
图像数据库 ·· 082

5.1 引言 ·· 082
  5.1.1 无人机图像处理系统 ·························· 083
5.2 航空影像数据库框架 ································ 084
  5.2.1 数据库要求 ······································ 085
  5.2.2 数据库设计 ······································ 085
5.3 图像捕获过程 ·· 085
5.4 结果 ·· 088
  5.4.1 收集的图像 ······································ 088
5.5 图像数据库的使用 ·································· 091
  5.5.1 图像镶嵌 ·· 091
  5.5.2 CV 算法的发展 ································· 093
5.6 结论与未来的工作 ·································· 095
参考文献 ·· 097

第6章 通信需求、视频传输、通信链路和网络化无人机 ······ 101

6.1 引言 ·· 101
6.2 飞行自组织网络 ····································· 102
6.3 FANET 协议 ·· 103
6.4 FANET:流媒体和监控 ····························· 106
6.5 讨论和未来趋势 ····································· 108
  6.5.1 FN 的布局搜索算法 ·························· 108
  6.5.2 事件检测和视频质量选择算法 ··············· 109
  6.5.3 机载视频管理(无人机) ······················ 110
  6.5.4 舰队平台视频速率适配 ······················ 110
  6.5.5 飞行节点协调 ·································· 110
  6.5.6 数据收集和显示 ······························· 111
  6.5.7 软件定义的网络 ······························· 111
  6.5.8 网络功能虚拟化 ······························· 112
  6.5.9 数据收集与能量收集 ························· 112
6.6 小结 ·· 114
参考文献 ·· 115

XVII

第7章 无人机多光谱与高光谱成像：现状与展望 ············ 119
7.1 引言 ············ 119
7.2 UAV成像体系架构和组成 ············ 121
7.2.1 UAV的未来应用领域 ············ 123
7.3 多光谱与高光谱成像仪 ············ 123
7.3.1 多光谱成像 ············ 123
7.3.2 高光谱成像 ············ 124
7.3.3 卫星成像与UAV成像 ············ 126
7.4 UAV图像处理工作流程 ············ 126
7.4.1 大气校正 ············ 126
7.4.2 光谱影响映射 ············ 127
7.4.3 降维 ············ 127
7.4.4 计算任务 ············ 128
7.5 空间数据的数据处理工具包 ············ 128
7.6 UAV开放式多光谱与高光谱数据集研究 ············ 131
7.7 MSI和HIS UAV成像应用 ············ 131
7.7.1 农业监测 ············ 132
7.7.2 沿海监视 ············ 132
7.7.3 林业 ············ 132
7.7.4 城市规划 ············ 132
7.7.5 国防应用 ············ 132
7.7.6 环境监测 ············ 133
7.7.7 其他商业用途 ············ 133
7.8 结论和未来展望 ············ 133
参考文献 ············ 134

第8章 无人机航空成像与基础设施重构 ············ 140
8.1 引言 ············ 140
8.2 相关研究 ············ 141
8.3 视觉传感器和任务规划器 ············ 143
8.3.1 图像投影 ············ 143
8.3.2 路径规划器 ············ 144
8.4 三维重构 ············ 144
8.4.1 立体映射 ············ 145
8.4.2 单目映射 ············ 146

8.5 数据集收集 ……………………………………………………… 147
   8.5.1 试验装置 …………………………………………………… 147
   8.5.2 数据集1 ……………………………………………………… 149
   8.5.3 数据集2 ……………………………………………………… 149
8.6 试验结果 ……………………………………………………… 150
   8.6.1 室内场景 …………………………………………………… 150
   8.6.2 室外场景1 …………………………………………………… 151
   8.6.3 室外场景2 …………………………………………………… 152
   8.6.4 地下场景 …………………………………………………… 153
8.7 未来的发展趋势 ………………………………………………… 154
8.8 小结 …………………………………………………………… 154
参考文献 ……………………………………………………………… 155

第9章 深度学习作为无人机系统中超分辨率成像的可替代方案 … 159

9.1 引言 …………………………………………………………… 159
9.2 超分辨率模型 …………………………………………………… 160
   9.2.1 运动估计 …………………………………………………… 162
   9.2.2 去模糊 ……………………………………………………… 164
   9.2.3 斑块选择 …………………………………………………… 164
   9.2.4 超分辨率 …………………………………………………… 164
9.3 试验和结果 ……………………………………………………… 166
   9.3.1 峰值信噪比 ………………………………………………… 167
9.4 UAV-CPS中超分辨率运用的关键问题 ………………………… 168
   9.4.1 大数据 ……………………………………………………… 168
   9.4.2 云计算服务 ………………………………………………… 168
   9.4.3 图像捕获硬件限制 ………………………………………… 169
   9.4.4 视频超分辨率 ……………………………………………… 170
   9.4.5 高效度量及其他评估策略 ………………………………… 171
   9.4.6 多重先验 …………………………………………………… 172
   9.4.7 正则化 ……………………………………………………… 172
   9.4.8 新颖体系架构 ……………………………………………… 173
   9.4.9 3D超分辨率 ………………………………………………… 175
   9.4.10 深度学习和计算智能 ……………………………………… 176
   9.4.11 网络设计 …………………………………………………… 178
9.5 小结 …………………………………………………………… 179
参考文献 ……………………………………………………………… 179

第 10 章　无人机系统的体验质量和服务质量 ·········· 193
10.1　引言 ·········· 193
10.1.1　从 CPS 角度看机载网络 ·········· 194
10.2　概述 ·········· 196
10.2.1　影响 QoS/QoE 的参数 ·········· 197
10.2.2　云距离对于 QoS/QoE 的影响 ·········· 197
10.2.3　UAV-CPS 中的 QoS/QoE 监视框架 ·········· 198
10.2.4　应用层管理 ·········· 200
10.2.5　网络级管理 ·········· 200
10.2.6　云距离管理 ·········· 200
10.2.7　QoS/QoE 服务级别管理 ·········· 200
10.2.8　UAV-CPS 中的 QoS/QoE 指标 ·········· 201
10.2.9　QoS 到 QoE 的映射 ·········· 201
10.2.10　主观与客观测量 ·········· 201
10.2.11　衡量 QoS/QoE 的工具 ·········· 202
10.3　应用领域 ·········· 203
10.3.1　社交网络、游戏和人机界面 ·········· 203
10.3.2　数据中心 ·········· 204
10.3.3　电网和能源系统 ·········· 204
10.3.4　网络系统 ·········· 204
10.3.5　监视 ·········· 205
10.4　实例探究 ·········· 205
10.4.1　应用场景1：UAV-CPS 在交通拥堵管理中的应用 ·········· 205
10.4.2　应用场景2：使用智能车辆系统避免拥堵和事故 ·········· 207
10.5　未来与挑战 ·········· 209
10.5.1　建模和设计 ·········· 209
10.5.2　协作服务 ·········· 210
10.5.3　流 ·········· 210
10.5.4　安全 ·········· 211
10.5.5　飞行特设网络 ·········· 211
10.5.6　用户情绪 ·········· 213
10.6　小结 ·········· 213
参考文献 ·········· 214

第 11 章　总结与展望 ·········· 222

第1章 作为新技术测试台和工业5.0基础的无人机信息-物理系统

广泛的云资源和基础设施可以为无人机信息-物理系统(UAV-CPS)提供重大改进,UAV-CPS依赖于来自网络的数据或代码来操作并不全部来自单一固定结构的传感器、执行器、计算模块和存储仓库。本章围绕着云的潜在优势进行组织:① 对可视化资料库的大数据(BD)访问,其中包含表示/描述性数据、图像、视频、地图和飞行路径;② 面向网格计算(GC)的云计算功能,按需满足统计分析、机器学习(ML)算法、计算智能(CI)应用和飞行规划的要求;③ 集体无人机学习(CUL),其中无人机共享其轨迹、控制指南和任务结果;④ 通过众包方式的人机协作,用于分析高维高分辨率(HDHR)图像/视频、场景/对象/实体分类、学习和纠错/隐藏。云也可以扩展 UAV-CPS,如提供:① 数据集、模型、各种文献、HDHR 基准和软件/硬件模拟器;② 针对开源硬件(OSH))的开放竞争;③ 开源软件(OSS)。本章讨论在 UAV-CPS 中的一些开放挑战和新趋势。

1.1 引言

云计算(CC)可以将无人机(UAV)建模为信息-物理系统(CPS)的一部分,从而扩大其应用范围和潜力[1]。这种方法非常有用,因为用户可以专注于他们的目标,而不用受维护、断电、软件/硬件更新、人群的存在和天气状况等因素的干扰。云还允许不同的处理规模,并简化了应用程序和用户之间的信息共享。大多数无人机网络-物理系统仍然只使用机载处理、内存和软件来工作。云中网络的不断发展和处理需要新的策略,即远程访问动态和有效的全球服务、手段和数据集,为各种功能和操作场景提供支持和资源的可扩展性[2]。

云计算(CC)使当前和即将推出的 UAV-CPS 能够使用无线网络服务、大数据(BD)、开源软件(OSS)、开源硬件(OSH)、网格计算(GC)、统计机器学习(ML)、计算智能(CI)、众包和其他集体资源,从而在各种用途中提升性能,如装配线、检验、驾驶、包裹运输、仓储物流、对残疾人和老年人的护理、防灾/减灾、天文学以及医疗应用。

UAV-CPS 系统在很大程度上依赖于机器人、操控员、终端用户、地面控制站(GCS)和通过有线和无线连接的遥感(RS)设备。由于网络延迟、灵活的服务质量(QoS)和停机时间,云 UAV-CPS 通常具有一些本地处理能力,用于低延迟响应以及在网络访问不存在或不可靠的整个期间。本章讨论了 CC 的潜在优势[3-12]如下。

(1)BD(大数据)访问图像、地图、轨迹和对象数据的远程库。

(2)CC(云计算)按需提供并行 GC(网格计算),用于视觉统计分析、学习,以及飞行规划。

(3)由于无人机共享控制策略、轨迹和结果而导致的集体无人机学习(CUL)。

(4)利用众包访问远程人类专业知识,用于分析图像、分类、学习和错误恢复。

本章还列举了云可以促进 UAV-CPS 的实例,可方便地访问:① 仿真工具、数据集、模型、基准和出版物;② 设计和系统的公开竞赛;③ OSS 之类的操作系统[13-19]。

基于云的 UAV-CPS 的概念允许其他一些新举措,以便"轻松对话",如机器人操作系统(ROS)[14]和物联网(IoT)。例如,RFID 模块和廉价的 FPGA 处理器可以在大量的 UAV、GCS 和来自医疗机构的物理对象中发挥多种作用,以实现通信、控制和信息交换。

云使得 UAV-CPS 能够访问大量数据资源以及不适合机载内存的信息存储库。BD(大数据)包含的数据超过了传统数据库系统的处理能力,包括图像、视频、RS 数据、地图、实时网络和商业交易以及大量传感器网络。

基于云的 UAV-CPS 的许多来源都可以收集到数据,其中 UAV-CPS 系统结合了来自分布式处理和多个传感器、无人机、地图和 RS 存储库的数据流,并考虑了 CC(云计算)的 BD(大数据)和 CUL(无人机集体学习),如图 1.1 所示。

一些 BD 算法的机遇和挑战可以为大型数据集的查询提供合理的预估,从而保证运行时间的可管理性,但这些近似值可能会受到退化数据的严重影响。

基于云的 UAV-CPS 可以使用来自移动设备的全球定位系统(GPS)来收集运输数据、处理它们并分发命令,同时收集和共享多种类型的环境参数信息,如噪声和污染等级。

大数据集可以促进机器学习,这在计算机视觉(CV)领域中已经得到了证

第1章 作为新技术测试台和工业5.0基础的无人机信息-物理系统

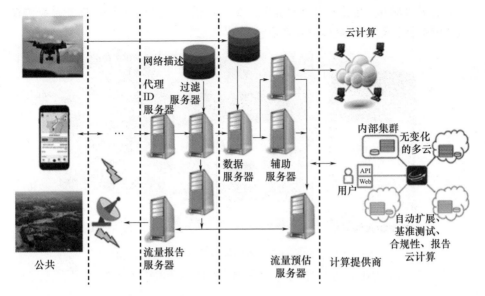

图1.1 基于云的UAV-CPS

明。大规模图像数据集已用于对象识别和场景识别,同时减少了手动标记训练数据的必要性。云处理采用开放的可视化数据集,提高了增强现实(AR)应用程序的有效性。将互联网图像与本地人工的可视化语义查询相结合,可以提供更强大的CV(计算机视觉)学习算法[20-27]。深度学习(DL)依赖于多层神经网络,而神经网络获益于BD(大数据),并应用于CV(计算机视觉)[28-29]。

导航是UAV-CPS面临的一个持久性挑战,除了在线数据库中针对三维或四维模型的演示外,还可以通过匹配二维图像结构、三维特征、三维点云的传感器信息,从依赖增量学习的云资源中获益[30-31]。谷歌地图可以改善基于云的UAV-CPS[32],如下面所述。

无人机捕获图像并通过网络将其发送到对象识别服务器(ORS),该服务器将数据返还给候选项集,每个候选项集都有预先计算的评估选项。无人机将返回的CAD模型与检测到的点云进行比较,以优化识别、估计姿态并选择适当的实施方案。例如,在导航之后,当前路径和状态上的数据将用于更新云中模型,以供将来参考[33-34]。未来可以使用云支持的自动化BD(大数据)、CC(云计算)和CUL(无人机集体学习)。

UAV-CPS存储与任何类型的对象/实体及地图相关的数据,用于目标识别、移动导航、抓取和操纵等(图1.2,图1.3)[36-37]。由多个在线数据集来评估不同算法的性能。

UAV-CPS面临的一项研究挑战是如何定义用于表示数据的跨平台格式。传感器获取的数据(如图像、视频和点云)格式很少能广泛使用。即使是相对简单的数据,如轨迹,也没有通用的标准。另一个需要注意的问题是对于有效传输数据的稀疏表示,如稀疏运动规划算法[38-41]。

图 1.2　谷歌的对象识别系统将大型视觉数据集和
文本标签(元数据)与 ML 相结合,以简化云中的对象识别[35]

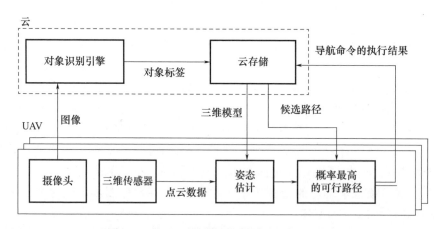

图 1.3　基于云的对象识别服务器(ORS)系统

从分布式数据源收集的大型数据集定期将损坏的信息与错误、重复或降级的数据(如在无人机校准过程中收集的三维位置数据[44])封装在一起[42,43]。新方法必须对未经处理的原始数据具有稳健性[45-49]。

1.2　云计算

大规模的按需并行计算广泛应用于商业中[50],如亚马逊的弹性计算云、微软的 Azure[51] 和谷歌的计算引擎。这些系统为短期任务提供对大量的远程计算机

第 1 章 作为新技术测试台和工业 5.0 基础的无人机信息 – 物理系统

访问。这些服务最初主要由 Web 应用程序开发人员使用,如今在科学和技术的高性能计算(HPC)应用程序中越来越流行[51-56]。

感知、模型和控制的不确定性是 UAV – CPS 的核心问题,它们可以建模为位置、形状、方向和控制方面的扰动。CC(云计算)是基于样本的蒙特卡罗分析的理想选择[57]。例如,并行云计算可用于计算物体/环境姿态、形状以及无人机对传感器和命令的响应中许多可能扰动的叉积结果。医学和粒子物理学正探索这个想法。

当存在形状不确定性时,基于云的采样有助于计算稳健轨迹。例如,飞行规划算法可以接受每个顶点和质心周围遵循高斯分布的多边形轮廓作为输入,并利用并行采样的方法来计算多个任务的质量指标。

云计算有可能加速许多计算密集型 UAV – CPS 任务,如图 1.4 所示的云中 SLAM 导航,下一视点规划针对目标识别和无人机编队控制。

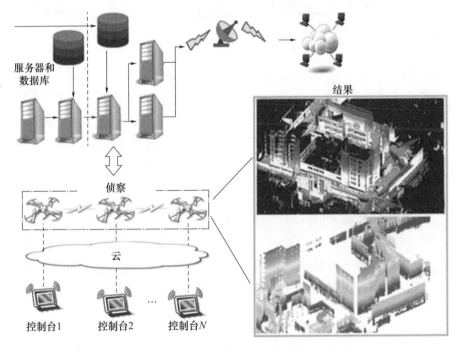

图 1.4 使用云内协作跟踪和映射(C2TAM)的无人机导航[58]

对于基于最佳采样的路径规划方法[36,59],云计算可用于生成图形;同样重要的是要认识到,这些图表数量很大。因此,图简化算法对于简化数据传输是必要的,如图 1.3 所示。

云还有助于可视化数据分析[60,61]、地图绘制[62](图 1.2,图 1.3)和辅助生活技术[63,64]。

图1.4演示了如何在云中使用同步定位与建图(SLAM)进行导航的计算机密集型光束平差法[58,65]。

图1.5展示了高维高分辨率(HDHR)空间中的二维运动规划图,以节省计算和网络资源。有效的无人机运动规划可以使操作更灵活、更简单,并且只有在需要时,无人机和云才能够分开计算。

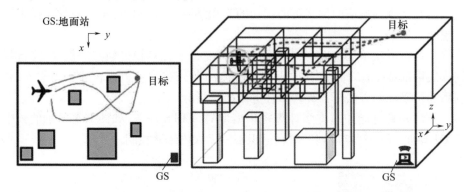

图1.5 基于分布式采样的飞行规划

认识到云容易出现波动的网络延迟和QoS是至关重要的。一些方法对时间不敏感,如整理或预计算导航、机器调度的离线优化。另外,许多任务包含实时需求,这是一个正在研究的领域。各种UAV–CPS的自驱动传感系统的开放标准也有待提高。使用云辅助储存来自传感器和执行器的网络数据,可以在考虑CUL的同时,实现运输路线和其他应用程序信息的协同共享[66-67]。

1.3 集体无人机学习

云通过收集多个实际试验和环境实例,简化了无人机的数据共享和学习。例如,UAV–CPS子系统可以共享初始状态和查找条件、配套控制策略和轨迹,以及有关性能及结果的重要数据。

图1.6显示了一个CUL框架,该框架可以索引许多无人机在无数任务中的轨迹,并使用CC进行并行规划和轨迹调整[68]。UAV–CPS可以学习云中存储的预先计算的飞行方案。规划器尝试找到一个新方案,并找出有相似问题的现有方案。先完成的方案被选中,同时这个框架可以使用云机器人技术和自动化的BD、CC和CUL。

根据图1.5和图1.6所示[33-34],此类系统还可以扩展到全球网络,以协助共享路径规划,包括业务路由选定。

第 1 章　作为新技术测试台和工业 5.0 基础的无人机信息-物理系统

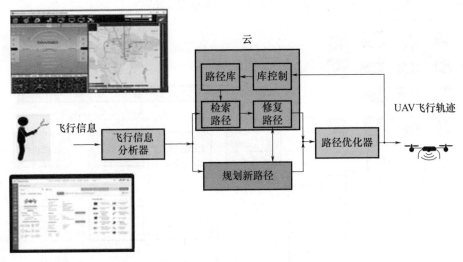

图 1.6　路径规划的架构示意图

通过 CUL 数据共享还可以提高计算资源受限的无人机能力。无人机可以像人类一样从社交、协作和共享中受益。此外，无人机还可以从共享传感器材料中受益，通过传感器得知其当前状态。

UAV-CPS 的数据库必须更新所连接的无人机以及其他来源的新案例，以便从公共互联网资源、计算机仿真和实际案例中学习。

1.4　人类计算、众包和呼叫中心

人类的技能、经验和直觉用来解决一些特殊情景中问题，如 CV 的图像标记、类标签和位置之间的关联规则以及数据采集。不久，无人机将按需在众包的基础上进行分配和调度，经历过再分配的人工可以执行超出计算机能力的任务。与自动预订系统相反，未来将考虑无人机和分布式计算机系统感知到的错误和异常，然后这些系统会联系远程呼叫中心的人类或高级人工智能单位寻求指导。

研究项目探讨了如何将其应用于路径规划、确定图像的深度层、法线和对称性，此外，还可以优化图像分割。研究人员正在努力了解定价模型，并利用基于知识的解决方案进行众包导航。

网络计算使得无人机可以远程操作，而扩展的云资源也促进了对远程人工操作的新研究。

图 1.7 描述了操控员使用一组不同策略控制 UAV-CPS 任务的界面。结果表明，人类可以选择更好、更稳健的策略。众包对象识别通过云 UAV-CPS 帮助

无人机导航,它将关于世界的语义知识与主观决策相结合。

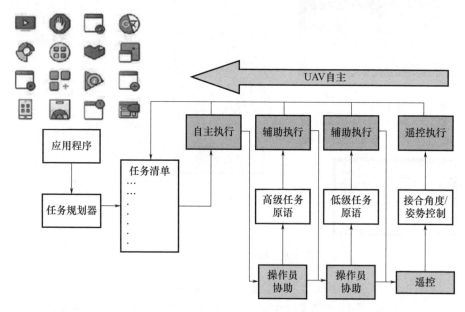

图 1.7 基于云的远程操作的人工协助

DARPA 机器人挑战赛(DRC)使用开源的基于云的仿真平台,对各种灾害管理任务进行 UAV – CPS 性能测试。云允许运行协作、实时并行仿真(如预测和评估性能、确认设计决策、优化设计、培训用户等)。

1.5 开源和开放访问的资源

云能够使人类顺利访问:① 数据集、出版物、模型、基准和仿真工具;② 开源硬件设计;③ 开放源码软件。并以此维持云支持无人机 CP 的持续发展。

OSS 的成功是显而易见的,如 ROS[13-14],它为软件开发者提供了库和工具。

ROS 也适用于 Android,因此被学术和工业领域的无人机开发人员广泛使用。

此外,一些 UAV – CPS 仿真库现在是 OSS,它允许个人快速建立和适应新系统,并共享生成的软件。还有许多开源仿真库、视频游戏模拟器、专门针对 UAV – CPS 的仿真环境、运动规划库和其他任务。这些 OSS 和 OSH 库允许修改,以适应应用程序和与原始设计不兼容的地方。

图 1.8 展示了一架由可回收材料制成的四轴飞行器,它可以在贫穷国家发挥作用,在这些国家,无人机可以利用剩余或回收的零部件[1,69]。

图1.8　四轴飞行器:由电子垃圾和再制造零件组成的无人机[69]

另一个发展趋势是在开源硬件领域,CAD模型和技术设备的细节可以自由获得。Arduino是一个使用广泛的OSH平台,它提供了多种类型的传感器和执行器。开放式医疗无人机的发展也需要特别关注。三维打印的最新进展将对无数领域产生重大影响,包括人道主义工程设计的OSH开发[70]。

云支持开放式设计挑战,吸引了各地创新者群体。

OSH可以促进云的使用,用于各种灾难任务的模拟和UAV-CPS性能测试。云支持交互式实时并行模拟,用于预测和评估性能、设计验证、优化和培训用户。

1.6　挑战和未来的方向

云固有的连接性除了带来挑战外,还带来了一系列新的隐私和安全问题。这些问题包括来自云连接的无人机、传感器和执行器的数据,特别是它们可能包含的HDHR多媒体数据[7,71-73]。无人机自动化服务也使得无人机可能遭受黑客攻击,使其功能受损或造成损失。例如,无人机可能受到廉价GPS欺骗方案的干扰。这些问题需要新的管理体制、问责机制,以及安全透明的法律法规。

为了解决网络时延和QoS随时间变化的问题,需要在技术层面上提出新的算法和方法。更快的数据网络、长期演进(LTE)[74]等有线和无线的互联网连接标准,减少了延迟。但是,当云资源长时间杂乱或不可用时,算法必须降级。例如,移动设备的负载均衡算法随时将信息发送到云端进行分析和处理,在合理延迟后使用最佳结果。

此外,还需要新的算法来扩展 BD 大小,这些算法通常包含未经处理的原始数据,需要进行充分的清理或采样。在分布式环境中,一些远程处理器在提交结果时可能会发生故障或经历长时间延迟,这一点对于算法过采样是至关重要的。在使用语义计算时,算法需要过滤不可靠的输入,同时平衡人工成本和无人机故障成本。

将 UAV-CPS 算法部署到云中需要简化过渡过程[75-81]。云架构可以分为三层。最低层是基础设施即服务(IaaS),提供在(可能是虚拟化的)机器上工作的裸操作系统。下一层,平台即服务(PaaS),提供更多结构,包括具有数据库访问权限的应用程序框架,同时限制可以采用的编程语言、架构和数据库模型的选择。最高的结构层称为软件即服务(SaaS),它的典型代表是基于云的谷歌 Docs 与需要下载后本地安装的 Microsoft Word 旧版本之间的区别。

无人机可以与云计算平台通信,云计算平台遵循 PaaS 范式,将计算从无人机转移到云端。它还连接知识库,进行大数据处理。这种 PaaS 原理可以扩展到 SaaS 范式,这为 UAV-CPS 提供了许多潜在的优势。使用 SaaS,接口允许将数据发送给处理它并返回输出的服务器,这不需要数据/硬件/软件维护,并允许公司控制专有软件。

这种方法可以称为机器人和自动化即服务(RAaaS)。为了解释这个概念,请参考使用 Microsoft Kinect RGBD 传感器设置 UAV 工作单元的两种情况。工作单元的目的是提取和检查装配线零件,需要对象识别/定位和飞行计划等。

ROS 有一个著名的开源机器人库,它提供对无数 OSS 包的访问,并在本地运行[14]。许多稳定的 ROS 包可以简化部署,但是有些软件只作为源代码分发提供,这需要下载和安装依赖项。操作人员必须设置一台新机器,如 Ubuntu,并解决所有库依赖项,包括与其他包的冲突。

相比之下,RAaaS 分析和规划软件在云中运行,用户访问网站,输入无人机、传感器、执行器和模型的输入数据和命令。然后,他们可以选择所需的识别和定位算法、飞行规划和其他智能算法,并利用人机界面将这些过程执行成流水线。无人机可以从 Kinect 等平台以点云的形式发送数据。无人机接收并执行运动计划和其他任务,并将结果反馈给基于云的 UAV-CPS,随着时间的推移改进云服务。

一些利用多媒体和高维数据(MHDD)特性的多媒体跨层优化(MCLO)方法用于传输速率调节、节能、错误恢复/隐藏、拥塞控制和多路径选择[82-87]。然而,资源受限的视觉传感器网络(VSN)[88-95]中的多媒体流问题在 UAV-CPS 中相当具有挑战性,这些问题尚未完全解决,新的并发症仍在出现。

在 UAV-CPS 系统中,引导未来 MHDD 研究的一个重要因素是方式表示技术和多媒体编码的进步。在 VSN 中,新的编解码器不断出现,在数据压缩、计算复杂性和错误恢复方面产生不同的结果。长期标准得到进一步完善,同时 VSN 已经采用了网络计算和分布式多媒体编码等新程序[45-49]。

关于 MHDD 的演变,了解哪些数据与单个应用程序相关是很有必要的。例如,具有地址事件表示的视频压缩可以以较低的计算成本实现帧差编码。在 VSN 中,帧间差分信息可以反映出错位目标的运动。在某些应用中,目标运动行为知识比视觉事实更重要。这样做时,可能会保留目标身份,但是旁观者仍然能够理解目标行为,因为即使与灰度图像相比,非常精简的代码也可以表征帧差数据。

MHDD 演进可以通过压缩感知(CS)和 DL 流进行跨层性能评估。与 JPEG 图像相比,由于图像表示缺乏结构性和充分性,CS 表示的图像显示出链接不准确的内在弹性。

随着计算模型、编码模型和感知模型的发展,虚拟存储网络中 MCLO 的部署将受到极大的推动。然而,诸如无线 VSN 的跟踪系统之类的其他事项需要适当地考虑,因为它们涉及多个移动接收器。除了多媒体流的编码方案外,还有一些问题可能改变路径选择协议、拥塞控制、错误修复。许多研究人员认为,带有超宽带(UWB)收发器的传感器可能增加成本。这种物理层技术允许的传输速率远高于 ZigBee[42,82-83,96-97]。有些人声称 UWB 是无线多媒体 VSN(WMVSN)的最终解决方案,MCLO 将不得不考虑 ad hoc VSN 的好处和流媒体挑战,其中 UWB 链路连接了中间节点[37,57,59,98-102]。

MAC 协议也可能影响基于多媒体的跨层(MCL)设计。例如,IEEE 802.11 标准可能更适合链路层,但 IEEE 802.15.4 仍然受欢迎。MAC 协议在未来实际的 VSN 应用中可能很少使用,这主要是受传感器节点的功率限制。尽管如此,还有一些用于 WMVSN 的能量感知 MAC 协议,如 T-MAC 和 ET-MAC。

无线链路的类型也可能影响跨层结构设计。总体而言,VSN 中的数据包丢失是网络瓶颈或位错误的结果。通过无线链路发送数据包,可能会在传输过程中产生误码。单个比特上的线性错误概率是不现实的,比特错误一般为突发且大数据包丢失的概率高于小数据包。由于减少的数据包意味着额外的协议开销,研究人员应该考虑传输多媒体编码信息的数据包的理想维数。一种改善方法是在数据包大小中使用 MCLO。然而,大量的科学研究将解决有关传输数据包的范围以及如何更有效地处理这些数据包的问题。缺少数据包大小的选择规范是明显且实际的原因。

由于功率限制仍然是未来几年的主要关注点,因此 VSN 中的 MCL 设计必须在实现所需优化的同时节省功率,或者至少不会产生额外的功耗[103-110]。许多文献作品提到 MCLO,却没有提倡能源意识。未来的研究可以增加节电机制,同时为 VSN 提供可行的解决方案。

与传输速率校正、能量节约、拥塞控制、错误恢复/隐藏、多路径广播和网络内压缩相关的新挑战仍然会出现,因此需要对 MCLO 进行更多的研究。一些可能的挑战来自特定监视实践的严格要求,如关于多层设计或独立来源。源节点移动性和覆盖保护等问题也需要学术界的创新研究,这直接影响到所采用的编码策略和实现的跨层部署。简而言之,在未来的工作中也应该考虑基于 CV 技术(如前景区分)建立的优先级方案。

1.7 小结

由于并非所有传感器/执行器部件、计算模块和内存仓库都具有等效的固定物理框架[20-27],当前广泛的云资源和基础可以通过学习或代码来改进 UAV – CPS。本章讨论了未来的云优势:

(1) BD 访问视觉库、基准视频、不同的记录、地图和方向。

(2) 并行网格计算的云计算功能,用于测量检查、ML 计算、CI 应用程序和飞行规划。

(3) CUL 无人机共享有关其方向、控制方法和结果的信息。

(4) 众包,研究 HDHR 图片/视频、安排、学习以及错误纠正和伪装。

云同样可以通过提供:① 数据集、编写、模型、测试系统和广泛的基准;② 利用 OSH 框架计划的公开竞争;③ OSS,这三种途径来发展 UAV – CPS。未来将考虑到 UAV – CPS 中的一些已知困难和新模式[111-114]。

MCLO 利用系统的多媒体编码内在特性,实现了更高水平的 VSN 设计。研究表明,在传输速率变化、功率管理/分配、拥塞控制、错误恢复、多路径决策和网络压缩等领域给予适当的投入,将推动新的创新实用程序发展并揭示新的需求。本章最后展望了未来的研究方向,指出了该领域的研究前景。新兴且逐渐成熟的 VSN 改善了 MCLO。

参考文献

[1] Estrela V. V. , Saotome O. , Loschi H. J. , et al. "Emergency Response Cyber Physical Framework for Landslide Avoidance with Sustainable Electronics," Technologies, vol. 6, no. 42, 2018. doi:10. 3390/technologies6020042.

[2] Kehoe B. , Patil S. , Abbeel P. , and Goldberg K. "A Survey of Research on Cloud Robotics and Automation," IEEE Transactions on Automation Science and Engineering, 2014;12(2):

第 1 章　作为新技术测试台和工业 5.0 基础的无人机信息－物理系统

398－409.

[3] Aguero C. ,Koenig N. ,Chen I. ,et al. "Inside the Virtual Robotics Challenge:Simulating Real-time Robotic Disaster Response," *IEEE Transactions on Automation Science and Engineering (T-ASE):SI on Cloud Robotics and Automation*, vol. 12,no. 2,2015,pp. 494－506.

[4] Bekris K. ,Shome R. ,Krontiris A. , and Dobson A. "Cloud Automation:Precomputing Roadmaps for Flexible Manipulation," *IEEE Robotics & Automation Magazine:SI on Emerging Advances and Applications in Automation*, 2014;22(2):41－50.

[5] Hu G. ,Tay W. ,and Wen Y. "Cloud Robotics:Architecture, Challenges and Applications," *IEEE Network*, vol. 26,no. 3,pp. 21－28,2012.

[6] Hunter T. ,Moldovan T. ,Zaharia M. ,et al. "Scaling the Mobile Millennium System in the Cloud," in *Proc. 2nd ACM Symposium on Cloud Comp (SOCC)*,Cascais Portugal,2011,p. 28.

[7] Jangid N. K. "Real Time Cloud Computing," in *Data Management & Security*, 2011.

[8] Mehrotra P. ,Djomehri J. ,Heistand S. ,et al. "Performance Evaluation of Amazon EC2 for NASA HPC Applications," in *Proc. 3rd ACM Workshop on Scientific Cloud Computing Data-Science Cloud'* 12,2012,New York,NY,USA,pp. 41－50.

[9] Michael N. ,Mellinger D. ,Lindsey Q. ,and Kumar V. "The GRASP Multiple Micro UAV Testbed," *IEEE Robotics & Automation Magazine*,vol. 17,no. 3,pp. 56－65,2010.

[10] Narita M. ,Okabe S. ,Kato Y. ,Murakwa Y. ,Okabayashi K. ,and Kanda S. "Reliable Cloud-Based Robot Services," in *Proc. Conference on the IEEE Industrial Electronics Society*. IEEE, 2013,Vienna,Austria,pp. 8317－8322.

[11] Nurmi D. ,Wolski R. ,Grzegorczyk C. ,et al. "The Eucalyptus Open-Source Cloud Computing System," in *Proc. IEEE/ACM International Symposium on Cluster Computing and the Grid*, Shanghai,China,2009,pp. 124－131.

[12] Proia A. A. ,Simshaw D. ,and Hauser K. "Consumer Cloud Robotics and the Fair Information Practice Principles:Recognizing the Challenges and Opportunities Ahead," *Minnesota Journal of Law,Science & Technology*,2015;16(1):1－70.

[13] Quigley M. ,Gerkey B. ,Conley K. ,et al. "ROS:An Open-Source Robot Operating System," in Proc. IEEE ICRA Workshop on Open Source Software,Kobe,Japan,2009.

[14] Estrela V. V. ,Hemanth, J. ,Nascimento, D. A. ,Loschi, H. J. ,and Razmjooy, N. "Computer Vision and Data Storage in UAVs," in *Advances in UAV Avionics for Imaging and Sensing*, vol. 1. IET,2019.

[15] Plaku E. ,Bekris K. E. ,and Kavraki L. E. "OOPS for Motion Planning:An Online,Open-Source,Programming System," in *Proc. IEEE International Conference on Robotics and Automation (ICRA)*, Rome,Italy,2007,pp. 3711－3716.

[16] Nurmi D. ,Wolski R. ,Grzegorczyk C. ,et al. "The Eucalyptus Open-Source Cloud Computing System," in Proc. *IEEE IEEE/ACM International Symposium on Cluster Computing and the Grid*,Shanghai,China,2009,pp. 124－131.

[17] Hannaford B. ,Rosen J. ,Friedman D. ,et al. "Raven II:Open Platform for Surgical Robotics Research," *IEEE Transactions on Biomedical Engineering*,60,pp. 954－959,2012.

[18] Davidson S. "Open‐Source Hardware," *IEEE Design and Test of Computers*, vol. 21, no. 5, pp. 456–456, 2004.

[19] Dabbish L., Stuart C., Tsay J., and Herbsleb J. "Social Coding in GitHub: Transparency and Collaboration in an Open Software Repository," in *Proc. IEEE ACM 2012 Conference on Computer Supported Cooperative Work*, Bellevue, Washington, USA, 2012, p. 1277.

[20] Gouveia B., Portugal D., Silva D., and Marques L. "Computation Sharing in Distributed Robotic Systems: A Case Study on SLAM," *IEEE Transactions on Automation Science and Engineering (T–ASE): SI on Cloud Robotics and Automation*, vol. 12, no. 2, 2015 pp. 10–22.

[21] Hidago‐Pena E., Marin‐Urias L. F., Montes‐Gonzalez F., Marin‐Hernandez A., and Rios‐Figueroa H. V., "Learning from the Web: Recognition Method Based on Object Appearance from Internet Images," in *Proc. ACM/IEEE International Conference on Human‐Robot Interaction*. IEEE, Tokyo, Japan, 2013, pp. 139–140.

[22] Krizhevsky A., Sutskever I., and Hinton G. E. "ImageNet Classification with Deep Convolutional Neural Networks," in *Advances in Neural Information Processing Systems* 25, Pereira F., Burges C., Bottou L., and Weinberger K., Eds. Curran Associates, Inc., 2012, pp. 1097–1105.

[23] Johnson‐Roberson M., Bohg J., Skantze G., et al. "Enhanced Visual Scene Understanding through Human‐Robot Dialog," in *Proc. IEEE International Conference on Intelligent Robots and Systems (IROS)*, San Francisco, CA, USA, 2011, pp. 3342–3348.

[24] Le Q., Ranzato M., Monga R., et al. "Building High‐Level Features Using Large Scale Unsupervised Learning," in *Proc. International Conference in Machine Learning*, ICML 2012, Edinburgh, Scotland, 2012.

[25] Samadi M., Kollar T., and Veloso M. "Using the Web to Interactively Learn to Find Objects," in *Proc. IEEE AAAI Conference on Artificial Intelligence*, 2012, Toronto, Ontario, Canada, pp. 2074–2080.

[26] Sevior M., Fifield T., and Katayama N., "Belle Monte‐Carlo Production on the Amazon EC2Cloud," *Journal of Physics: Conference Series*, vol. 219, no. 1, 2010.

[27] Chandrasekaran V., and Jordan M. I., "Computational and Statistical Tradeoffs Via Convex Relaxation," *Proc. National Academy of Sciences of the United States of America*, vol. 110, no. 13, pp. E1181–E1190, 2013.

[28] Hemanth D. J., and Estrela V. V. "Deep Learning for Image Processing Applications," *Advances in Parallel Computers*. IOS Press, 2017. ISBN 978‐1‐61499‐821‐1 (print) 978‐1‐61499‐822‐8 (online).

[29] Dean J., Corrado G., Monga R., et al. "Large Scale Distributed Deep Networks," in *Advances in Neural Information Processing Systems* 25, Pereira F., Burges C., Bottou L., and Weinberger K., Eds. Curran Associates, Inc., 2012, pp. 1223–1231.

[30] Beksi W., and Papanikolopoulos N., "Point Cloud Culling for Robot Vision Tasks under Communication Constraints," in *Proc. IEEE International Conference on Intelligent Robots and Systems (IROS)*, Chicago, Illinois, USA, 2014.

[31] Lai K. ,and Fox D. "Object Recognition in 3D Point Clouds Using Web Data and Domain Adaptation," *International Journal of Robotics Research*(*IJRR*),vol. 29,no. 8,pp. 1019 – 1037,2010.

[32] Kasper A. ,Xue Z. ,and Dillmann R. "The KIT Object Models Database:An Object Model Database for Object Recognition,Localization and Manipulation in Service Robotics," *International Journal of Robotics Research*(*IJRR*),vol. 31,no. 8,pp. 927 – 934,2012.

[33] Higuera J. ,Xu A. ,Shkurti F. ,and Dudek G. "Socially – Driven Collective Path Planning for Robot Missions," *Proc. IEEE Conference on Computer and Robot Vision*,pp. 417 – 424,Toronto,ON,Canada,2012.

[34] Berenson D. ,Abbeel P. ,and Goldberg K. "A Robot Path Planning Framework that Learns from Experience," in *Proc. IEEE International Conference on Robotics and Automation*(*ICRA*),Minneapolis,MN,USA,2012,pp. 3671 – 3678.

[35] Kuffner J. "Cloud – Enabled Robots," in *Proc. IEEE – RAS International Conference on Humanoid Robots*,Nashville,TN,USA,2010.

[36] Marinho C. E. V. ,Estrela,V. V. ,Loschi,H. J. et al. "A Model for Medical Staff Idleness Minimization," in *Proceedings of the 4th Brazilian Technology Symposium*(*BTSym'18*). BTSym 2018. *Smart Innovation*,*Systems and Technologies*,vol. 140,Iano Y. ,Arthur R. ,Saotome O. ,Estrela V. V. ,and Loschi H. ,Eds. Springer,Cham,Campinas,SP,Brazil,2019.

[37] Padilha R. ,Iano Y. ,Monteiro A. C. B. ,Arthur R. ,and Estrela V. V. "Betterment Proposal to Multipath Fading Channels Potential to MIMO Systems," in *Proc. BTSym 2018. Smart Innovation*,*Systems and Technologies*,vol. 140,Iano Y. ,Arthur R. ,Saotome O. ,Estrela V. V. ,and Loschi H. ,Eds. Springer,Cham,Campinas,SP,Brazil,2019.

[38] Chilamkurti N. ,Zeadally S. ,Soni R. ,and Giambene G. "Wireless Multimedia Delivery over 802. 11e with Cross – Layer Optimization Techniques," *Multimedia Tools and Applications*,vol. 47,pp. 189 – 2051,2010.

[39] Li Z. ,O'Brien L. ,Zhang H. ,and Cai R,"On the Conceptualization of Performance Evaluation of IaaS Services," *IEEE Transactions on Services Computing*, vol. X, no. X, pp. 1 – 1,2014.

[40] Wu Y. ,Lim J. ,and Yang M. "Object Tracking Benchmark. " *IEEE Transactions on Pattern Analysis and Machine Intelligence*, vol. 37,pp. 1834 – 1848,2015.

[41] Marins H. R. ,and Estrela V. V. "On the Use of Motion Vectors for 2D and 3D Error Concealment in H. 264/AVC Video. Feature Detectors and Motion Detection in Video Processing," *IGI Global*, Hershey, PA, USA, 1, pp. 164 – 186, 2017. doi:10. 4018/978 – 1 – 5225 – 1025 – 3. ch008

[42] Almalkawi I. ,Zapata M. ,Al – Karaki J. ,and Morillo – Pozo J. "Wireless Multimedia Sensor Networks:Current Trends and Future Directions," *Sensors* vol. 10,pp. 6662 – 6717,2010.

[43] Wang L. ,Liu M. ,and Meng M. Q. – H,"Real – Time Multi – Sensor Data Retrieval for Cloud Robotic Systems," *IEEE Transactions on Automation Science and Engineering*(*T – ASE*):*Special Issue on Cloud Robotics and Automation*, vol. 12,no. 2,2015,pp. 507 – 518.

[44] Mahler J., Krishnan S., Laskey M., et al. "Learning Accurate Kinematic Control of Cable-Driven Surgical Robots Using Data Cleaning and Gaussian Process Regression," in *Proc. IEEE International Conference on Automation Science and Engineering (CASE)*, New Taipei, Taiwan, 2014.

[45] Ciocarlie M., Pantofaru C., Hsiao K., Bradski G., Brook P., and Dreyfuss E. "A Side of Data With My Robot," *IEEE Robotics & Automation Magazine*, vol. 18, no. 2, pp. 44–57, 2011.

[46] Du Z., Yang W., Chen Y., Sun X., Wang X., and Xu C. "Design of a Robot Cloud Center," in Proc. *International Symposium on Autonomous Decentralized Systems*, Tokyo & Hiroshima, Japan, 2011, pp. 269–275.

[47] de Croon G., Gerke P., and Sprinkhuizen-Kuyper I. "Crowdsourcing as a Methodology to Obtain Large and Varied Robotic Data Sets," in *Proc. IEEE International Conference on Intelligent Robots and Systems (IROS)*, Chicago, Illinois, USA, 2014.

[48] Deng J., Dong W., Socher R., Li L.-J., Li K., and Fei-Fei L. "Imagenet: A Large-Scale Hierarchical Image Database," in Proc. *IEEE Conference on Computer Vision and Pattern Recognition*, Miami, FL, USA, 2009, pp. 248–255.

[49] Wang J., Krishnan S., Franklin M. J., Goldberg K., Kraska T., and Milo T. "A Sample-and-Clean Framework for Fast and Accurate Query Processing on Dirty Data," in *Proc. ACM SIGMOD International Conference on Management of Data*, Snowbird, Utah, USA, 2014.

[50] Armbrust M., Stoica I., Zaharia M., et al. "A View of Cloud Computing," *Communications of the ACM*, vol. 53, no. 4, p. 50, 2010.

[51] Tudoran R., Costan A., Antoniu G., and Bouge L. "A Performance Evaluation of Azure and Nimbus Clouds for Scientific Applications," in *Proc. International Workshop on Cloud Computing Platforms-CloudCP' 12*. ACM Press, Bern, Switzerland, 2012, pp. 1–6.

[52] Tenorth M., and Beetz M. "KnowRob: A Knowledge Processing Infrastructure for Cognition-Enabled Robots," *International Journal of Robotics Research (IJRR)*, vol. 32, no. 5, pp. 566–590, 2013.

[53] Tenorth M., Clifford Perzylo A., Lafrenz R., and Beetz M. "The RoboEarth Language: Representing and Exchanging Knowledge About Actions, Objects, and Environments," in *Proc. International Conference on Robotics and Automation (ICRA)*, Minneapolis, MN, USA, 2012, pp. 1284–1289.

[54] Tenorth M., Perzylo A. C., Lafrenz R., and Beetz M. "Representation and Exchange of Knowledge about Actions, Objects, and Environments in the RoboEarth Framework," *IEEE Trans Aut Science and Eng (T-ASE)*, vol. 10, no. 3, pp. 643–651, 2013.

[55] Torralba A., Fergus R., and Freeman W. T. "80 Million Tiny Images: A Large Data Set for Nonparametric Object and Scene Recognition," *IEEE Transactions on Pattern Analysis and Machine Intelligence*, vol. 30, no. 11, pp. 1958–1970, 2008.

[56] Turnbull L., and Samanta B. "Cloud Robotics: Formation Control of a Multi Robot System Utilizing Cloud Infrastructure," in *Proceedings of IEEE Southeastcon*. IEEE, 2013, Jacksonville, FL, USA, pp. 1–4.

[57] Wang H., Ma Y., Pratx G., and Xing L. "Toward Real-time Monte Carlo Simulation Using a Commercial Cloud Computing Infrastructure," *Physics in Medicine and Biology*, vol. 56, no. 17, pp. N175–N181, 2011.

[58] Riazuelo L, Civera J., and Montiel J. "C2TAM: A Cloud Framework for Cooperative Tracking and Mapping," *Robotics and Autonomous Systems*, vol. 62, no. 4, pp. 401–413, 2013.

[59] van den Berg J., Abbeel P., and Goldberg K. "LQG-MP: Optimized Path Planning for Robots with Motion Uncertainty and Imperfect State Information," *International Journal on Robotics Research (IJRR)*, vol. 30, no. 7, pp. 895–913, 2011.

[60] Salmeron-Garcia J., Diaz-del Rio F., Inigo-Blasco P., and Cagigas D. "A Trade-off Analysis of a Cloud-based Robot Navigation Assistant using Stereo Image Processing," *IEEE Transactions on Automation Science and Engineering (T-ASE): SI on Cloud Robotics and Automation*, vol. 12, no. 2, 2015, pp. 444–454.

[61] Nister D., and Stewenius H. "Scalable Recognition with a Vocabulary Tree," in *Proc. IEEE Conference on Computer Vision and Pattern Recognition (CVPR)*, vol. 2, 2006, New York, NY, USA, pp. 2161–2168.

[62] Prestes E., Carbonera J. L., Fiorini S. R., et al. "Towards a Core Ontology for Robotics and Automation," *Robotics and Autonomous Systems*, vol. 61, no. 11, pp. 1193–1204, 2013.

[63] Khanna Y. E., Kim H., Jha S., and Parashar M. "Exploring the performance fluctuations of HPC workloads on clouds." Proc. 2010 IEEE Second International Conference on Cloud Computing Technology and Science, Indianapolis, IN, USA, 2010, pp. 383–387.

[64] Bhargava B., Angin P., and Duan L. "A Mobile-Cloud Pedestrian Crossing Guide for the Blind," in *Proc. International Conference on Advances in Computing & Communication*, Kochi, Kerala, India, 2011.

[65] Riazuelo L., Tenorth M., Marco D., et al. "RoboEarth Semantic Mapping: A Cloud Enabled Knowledge-Based Approach," *IEEE Transactions on Automation Science and Engineering (T-ASE): Special Issue on Cloud Robotics and Automation*, vol. 12, no. 2, 2015, pp. 432–443.

[66] Mohanarajah G., Hunziker D., Waibel M., and D'Andrea R. "Rapyuta: A Cloud Robotics Platform," *IEEE Transactions on Automation Science and Engineering (T-ASE)*, 12, pp. 1–13, 2014.

[67] Mohanarajah G., Usenko V., Singh M., Waibel M., and D'Andrea R. "Cloud-based Collaborative 3D Mapping in Real-Time with Low-Cost Robots," *IEEE Transactions on Automation Science and Engineering (T-ASE): Special Issue on Cloud Robotics and Automation*, vol. 12, no. 2, 2015, pp. 423–431.

[68] Kumar V., and Michael N. "Opportunities and Challenges with Autonomous Micro Aerial Vehicles," *International Journal of Robotics Research (IJRR)*, vol. 31, no. 11, pp. 1279–1291, 2012.

[69] https://www.instructables.com/id/Build-Your-Own-Eco-Quad-Copter-from-Reused-and-Rec/

[70] Lipson H., and Kurman M. *Fabricated: The New World of 3D Printing*. Wiley, 2013.

[71] Winkler T., and Rinner B. "Security and Privacy Protection in Visual Sensor Networks: A Survey," *ACM ComputingSurveys*, vol. 47, no. 2, pp. 1 – 2, 42, 2014.

[72] Ren K., Wang C., Wang Q., et al. "Security Challenges for the Public Cloud," *IEEE Internet Computing*, vol. 16, no. 1, pp. 69 – 73, 2012.

[73] Lin P., Abney K., and Bekey G. A., *Robot Ethics: The Ethical and Social Implications of Robotics.* The MIT Press, 2011.

[74] Astely D., Dahlman E., Furuskar A., Jading Y., Lindstrom M., and Parkvall S. "LTE: The Evolution of Mobile Broadband," *IEEE Communications Magazine*, vol. 47, no. 4, pp. 44 – 51, 2009.

[75] Arumugam R., Enti V., Bingbing L., et al. "DAvinCi: A Cloud Computing Framework for Service Robots," in *Proc. International Conference on Robotics and Automation (ICRA)*, Anchorage, AK, USA, 2010, pp. 3084 – 3089.

[76] Li Z., Zhang H., O'Brien L., Cai R., and Flint S. "On Evaluating Commercial Cloud Services: A Systematic Review," *Journal of Systems and Software*, vol. 86, no. 9, pp. 2371 – 2393, 2013.

[77] Gammeter S., Gassmann A., Bossard L., Quack T., and Van Gool L. "Server – Side Object Recognition and Client – Side Object Tracking for Mobile Augmented Reality," in *Proc. IEEE Conference on Computer Vision and Pattern Recognition*, 2010, San Francisco, CA, USA, pp. 1 – 8.

[78] Chow H. Y., Lui K. – S., and Lam E. Y. "Efficient On – Demand Image Transmission in Visual Sensor Networks," *EURASIP Journal on Advances in Signal Processing*, vol. 2007, pp. 1 – 11, 2007.

[79] Liu B., Chen Y., Blasch E., and Pham K. "A Holistic Cloud – Enabled Robotics System for Real – Time Video Tracking Application," in *Future Information Technology*, 2014, pp. 455 – 468.

[80] Juve G., Deelman E., Berriman G. B., Berman B. P., and Maechling P. "An Evaluation of the Cost and Performance of Scientific Workflows on Amazon EC2," *Journal of Grid Computing*, vol. 10, no. 1, pp. 5 – 21, 2012.

[81] Kamei K., Nishio S., Hagita N., and Sato M. "Cloud Networked Robotics," *IEEE Network*, vol. 26, no. 3, pp. 28 – 34, 2012.

[82] Campelli L., Akyildiz I. F., Fratta L., and Cesana M. "A Cross – Layer Solution for Ultrawideband Based Wireless Video Sensor Networks," in *Proceedings of the IEEE Global Telecommunications Conference*. New Orleans, LA, 2008, pp. 1 – 6.

[83] Khattak H. A., Ameer Z., Din I. U., and Khan M. K. "Cross – Layer Design and Optimization Techniques in Wireless Multimedia Sensor Networks for Smart Cities," *Computer Science and Information Systems*, 16, 1 – 17, 2019.

[84] Chen M., Leung V. C., Mao S., and Li M. "Cross – Layer and Path Priority Scheduling Based Real – time Video Communications over Wireless Sensor Networks," in *Proc. IEEE Vehicular Technology Conference*, Calgary, Canada, 2008, pp. 2873 – 2877.

[85] Lin C. – H., Shieh – K., Ke C. H., and Chilamkurti N. "An Adaptive Cross – Layer Mapping Algorithm for MPEG – 4 Video Transmission Over IEEE 802. 11e WLAN," *Telecommunication Systems*, vol. 42, pp. 223 – 234, 2009.

[86] Shu L., Zhang Y., Yu Z., Yang L. T., Hauswirth M., and Xiong N. "Context-Aware Cross-Layer Optimized Video Streaming in Wireless Multimedia Sensor Networks," *The Journal of Supercomputing*, vol. 54, pp. 94-121, 2010.

[87] Wan J., Xu X., Feng R., and Wu Y. "Cross-Layer Active Predictive Congestion Control Protocol for Wireless Sensor Networks," *Sensors*, vol. 9, pp. 8278-8310, 2009.

[88] Costa D. G., and Guedes L. A. "The Coverage Problem in Video-Based Wireless Sensor Networks: A Survey," *Sensors*, vol. 10, pp. 8215-8247, 2010.

[89] Chew L. W., Ang L.-M., and Seng K. P. "A Survey of Image Compression Algorithms in Wireless Sensor Networks," in *Proc. International Symposium on Information Technology*, 2008, Kuala Lumpur, Malaysia, pp. 1-9.

[90] Zhang L., Hauswirth M., Shu L., Zhou Z., Reynolds V., and Han G. "Multi-Priority Multi-Path Selection for Video Streaming in Wireless Multimedia Sensor Networks," *Lecture Notes in Computer Science*, vol. 5061, pp. 439-452, 2008.

[91] Lee J.-H. and Jun I.-B. "Adaptive-Compression Based Congestion Control Technique for Wireless Sensor Networks," *Sensors*, vol. 10, pp. 2919-2945, 2010.

[92] Wang H., Peng D., Wang W., and Sharif H. "Optimal Rate-based Image Transmissions via Multiple Paths in Wireless Sensor Network," in *Proc. IEEE Conference Multimedia and Expo*, Beijing, China, 2007, pp. 2146-2149.

[93] Nasri M., Helali A., Sghaier H., and Maaref H. "Adaptive Image Transfer for Wireless Sensor Networks (WSNs)," in *Proc. International Conference on Design and Technology of Integrated Systems in Nanoscale*, Hammamet, Tunisia, 2010, pp. 1-7.

[94] Qaisar S., and Radha H. "Multipath Multi-Stream Distributed Reliable Video Delivery in Wireless Sensor Networks," in *Proc. 43rd Annual Conference on Information Sciences and Systems (CISS 2009)*, 2009, pp. 207-212.

[95] Pudlewski S., and Melodia T. "On the Performance of Compressive Video Streaming for Wireless Multimedia Sensor Networks," in *Proc IEEE Ad-hoc, Sensor and Mesh Networking Symposium*, Shuzou, China, 2010, pp. 1-5.

[96] Wu F., Gui Y., Wang Z., Gao X., and Chen G. "A Survey on Barrier Coverage with Sensors," *Frontiers of Computer Science*, vol. 10, pp. 968-984, 2016.

[97] Akyildiz I., Melodia T., and Chowdhury K. "A Survey on Wireless Multimedia Sensor Networks," *Computer Networks*, vol. 51, pp. 921-960, 2007.

[98] Razmjooy N., Ramezani M., and Estrela V. V. "A Solution for Dubins Path Problem with Uncertainties Using World Cup Optimization and Chebyshev Polynomials," in *Proc. BTSym 2018. Smart Innovation, Systems and Technologies*, vol 140, Iano Y. et al. Eds. Springer, Cham, Campinas, SP, Brazil, 2019. doi: 10.1007/978-3-030-16053-1_5

[99] Turnbull L., and Samanta B. "Cloud Robotics: Formation Control of a Multi Robot System Utilizing Cloud Infrastructure," in *Proceedings of IEEE Southeastcon*. IEEE, 2013, Jacksonville, FL, USA, pp. 1-4.

[100] Waibel M., Beetz M., Civera J., *et al.* "RoboEarth," *IEEE Robotics & Automation Maga-

zine, vol. 18, no. 2, pp. 69 – 82, 2011.

[101] Razmjooy N., Ramezani M., Estrela V. V., Loschi H. J., and do Nascimento D. A. "Stability Analysis of the Interval Systems Based on Linear Matrix Inequalities," in *Proc. BTSym* 2018. *Smart Innovation, Systems and Technologies*, vol. 140, Iano Y. et al. Eds. Springer, Cham, Campinas, SP, Brazil, 2019.

[102] Razmjooy N., Razmjooy, N., Estrela, V. V., et al. "Image Transmission in MIMO – OSTBC System over Rayleigh Channel Using Multiple Description Coding (MDC) with QPSK Modulation," in *Advances in UAV Avionics for Imaging and Sensing*, vol. 2, IET, 2019.

[103] Mekonnen T., Porambage P., Harjula E., et al. "Energy Consumption Analysis of High Quality Multi – Tier Wireless Multimedia Sensor Network," *IEEE Access*, vol. 5, pp. 15848 – 15858, 2017.

[104] Yu W., Sahinoglu Z., and Vetro A. "Energy Efficient JPEG 2000 Image Transmission over Wireless Sensor Networks," in *Proceedings of the IEEE Globecom*, Dallas, Texas, USA, 2004, pp. 2738 – 2743.

[105] Lecuire V., Duran – Faundez C., and Krommenacker N. "Energy – Efficient Image Transmission in Sensor Networks," *International Journal of Sensor Networks*, vol. 4, pp. 37 – 47, 2008.

[106] Wang W., Peng D., Wang H., Sharif H., and Chen H. – H. "Energy – Constrained Distortion Reduction Optimization for Wavelet – Based Coded Image Transmission in Wireless Sensor Networks," *IEEE Transactions on Multimedia*, vol. 10, pp. 1169 – 1180, 2008.

[107] Wu H., and Abouzeid A. "Energy – Efficient Distributed Image Compression In Resource – Constrained Multihop Wireless Networks," *Computer Communications*, vol. 28, pp. 1658 – 1668, 2005.

[108] Aghdasi H. S., Abbaspour M., Moghadam M. E., and Samei Y. "An Energy – Efficient and High – Quality Video Transmission Architecture in Wireless Video – Based Sensor Networks," *Sensors*, vol. 8, pp. 4529 – 4559, 2008.

[109] Ferrigno L., Marano S., Pacielo V., and Pietrosanto A. "Balancing Computational and Transmission Power Consumption in Wireless Image Sensor Networks," in *Proc. 2005 IEEE Conference on Virtual Environments, Human – Computer Interfaces and Measurement Systems (VECIMS 2005)*, Messina, Italy, 2005, pp. 61 – 66.

[110] Politis I., Tsagkaropoulos M., Dagiuklas T., and Kotsopoulos S. "Power Efficient Video Multipath Transmission Over Wireless Multimedia Sensor Networks," *Mobile Networks and Applications*, vol. 13, pp. 274 – 284, 2008.

[111] Franc. a R. P., Peluso M., Monteiro A. C. B., Iano Y., Arthur R., and Estrela V. V. "Development of a Kernel: A Deeper Look at the Architecture of an Operating System," in *Proc. BTSym* 2018. *Smart Innovation, Systems and Technologies*, vol 140, Iano Y. et al. Eds. Springer, Cham, Campinas, SP, Brazil, 2019.

[112] Estrela V. V., Khelassi A., Monteiro A. C. B., et al. "Why software – defined radio (SDR) matters in healthcare?". Medical Technologies Journal, 2019; 3 (3): 421 – 429, doi: 10.

26415/2572-004X-vol3iss3p421-429.

[113] Estrela V. V., Hemanth J., Saotome O., Grata E. G. H., and Izario D. R. F. "Emergency response cyber-physical system for flood prevention with sustainable electronics." In: Iano Y. et al. (eds), Proceedings of the 3rd Brazilian Technology Symposium. BTSym 2017. Springer, Cham, Campinas, SP, Brazil, 2019.

[114] Stenmark M., Malec J., Nilsson K., and Robertsson A. "On Distributed Knowledge Bases for Small-Batch Assembly," *IEEE Transactions on Automation Science and Engineering (T-ASE): SI on Cloud Robotics and Automation*, vol. 12, no. 2, 2015.

第 2 章　无人机系统人为因素和人机界面设计

人机界面(HMI)是无人机系统(UAS)设计中的一个关键但又经常被忽视的方面。合理设计的人机界面不仅能增强态势感知能力,还能减少地面飞行员的工作量,从而有助于提高整体任务效能。通常,可利用人因工程(HFE)计划提供支持合理设计的方法过程。该计划分为三个迭代阶段:需求分析和捕获、设计和评估。在人因工程计划中可以采用一些方法,但鉴于不同类型无人机系统正在广泛地应用和赋予任务,采用功能性方法进行人机界面设计是有利的,即人机界面是围绕人类用户或系统要执行的特定功能来设计的。无人机系统典型功能包括任务规划、传感器操作、数据分析,以及感知与规避(SAA),还可扩展到多平台协调和协作决策。本章讨论了支持这些功能的人为因素考虑和相关的人机界面要素。

2.1　引言

无人机系统(UAS)具有更高的自动化和自主性,以弥补飞行员身体上从驾驶舱中解放出来的相关限制。与载人系统相比,UAS旨在实现更高的独立性和决策能力(如在制导、导航和控制(GNC)模块中[1])。无人机(UAV)的指挥和控制可以从更具战术性的模式(如手动驾驶)到更具战略性的模式变化,需要管理和协调多个载人和无人平台。因此,无人机系统人机界面和交互(HMI^2)的设计面临不同的挑战,尤其在涉及操作人员与自主代理之间协作的更复杂操作中。一些人因工程(HFE)考虑因素如下[2-5]。

(1)遥控站中不能获得驾驶舱内存在的某些感官(前庭、触觉、听觉)提示。

(2)数据链路的延迟和性能会限制地面飞行员向无人机提供战术或即时命令的能力。

(3)前面的限制可能会妨碍对异常运行情况的检测,以及随后从这种情况中的恢复。

(4)涉及长期低活动的任务,导致"人在环路之外"效应(如当执行高度自动化任务时),其特征是无聊感倍增、注意力下降和丧失态势感知能力。

(5)在不同飞行员/站之间转换无人机平台控制权或共享任务,特别是对于不同地理位置的飞行团队。

(6)由单个地面操控员管理多架无人机,要求低级别飞行任务自动化,以使地面操控员能够专注于更高级别的决策和协调任务。

(7)采用具备完整性监控和性能保证功能的系统建立可信的自主性。

(8)人机协作要求操控员将部分控制权或决策权让给自动或自主系统。无人机系统地面控制(GC)部分包含支持遥控机组人员控制和协调的一个或多个UAS平台的系统元素。无人机系统地面控制部分的尺寸各不相同,从小型移动单元到可容纳多名操控员的大型中心。按尺寸递增分为三种(单元、站、中心),以说明不同的操作规模。地面控制单元(GCU)由一名远程飞行员操作,设计为高度便携,包含遥控人员执行任务所有必要的功能和HMI[2],通常具有最小的外部依赖性。图2.1展示出了一款典型的地面控制单元,即一台笔记本电脑或手持设备,为地面飞行员提供视觉视线(VLOS)内无人机的战术控制。

图2.1 手持式地面控制单元示例

地面控制站(GCS)通常是一个由多个集成的硬件和软件组件的可部署架构。异构空基侦察小组(HART)地面控制站(图2.2)是一个支持地面部队的地面控制站示例,可实时地集成来自多个载人和无人平台的情报、监测和侦察信息。该HART地面控制站与指挥控制中心通信,并与其他联网地面控制站协调,以对任务和请求进行优先排序及分配。一个地面控制站通常配备多名操作人员,正如Peschel和Murphy[6]所述,操作人员承担的一般角色包括任务组织和计划、无人机驾驶和控制、传感器操作,以及数据处理、利用和传播。虽然当前地面控制站允许对单个无人机平台指挥和控制,但地面控制站的设计正在演变,以支持多平台操作,其中多个平台可以由多个操控员协调[7],从而允许执行更复杂的任务。

图2.2 异构空基侦察小组地面控制站示例,由诺斯罗普·格鲁曼公司提供

地面控制中心(GCC),如美国宇航局的全球鹰操作中心(图2.3),是一个固定结构,容纳多台计算机、显示器和控制器。地面控制中心将地面控制单元和地面控制站的元素集成在一个集中设施,允许执行复杂的任务,也可以支持多个无人机平台的协调。地面控制中心的特点是需要处理和传播大量数据、战略任务规划和高级别决策,以及在多个实体(如空中交通管理(ATM)以及其他中心、站和单元)之间协调计划和目标。

图2.3 地面控制中心示例——NASA全球鹰操作中心,图片由NASA提供

根据任务和操作需要,可重新配置和定制的地面控制显示器可以支持操作人员执行传统"飞行、导航、通信和管理"任务,该类任务也可在载人机中执行。地面控制界面可能包括与以下类别相关的显示和控制元素。

(1) 主要飞行状态。
(2) 导航和交通信息。
(3) 路径规划和任务管理。
(4) 系统状态信息和飞行器健康管理。
(5) 传感器馈电和有效载荷控制。
(6) 任务目标和任务进度。
(7) 通过文本或语音与其他操作人员通信。

2.2 无人机系统人机界面功能

随着自动化程度的提高,人工飞行任务被自动化功能所取代,使得遥控人员能够专注于管理和监控飞行和任务系统。表2.1提供了无人机系统机组任务的比较[6,8-10]。如表2.1所列,重点是管理任务的类型,分为三类:系统、数据和任务管理。数据和任务管理是无人机系统操作特有的任务,主要取决于任务和操作要求。数据管理与传感器数据的处理和利用有关,而任务管理与无人机系统资源的任务分派和分配有关。虽然民用飞行舱也包括数据和任务管理的某些方面,但它们并不被认为是飞行人员的主要任务之一。

表2.1 无人机系统"飞行、导航、通信和管理"任务

任务	Nehme 等[8]	Ashdown 等[9]	Peschel 和 Murphy[6]	Ramos[10]
飞行和导航	无人机位置的监控和优化	无人机操作	无人机远程操作导航	飞机指挥和控制
通信	与相关方协商并通知他们	通信	与其他团队通信	语音通信,有效载荷成果的传播散发
管理系统	监控无人机的健康和状态,监控网络通信,监控载荷状态	载具系统管理,有效载荷系统管理	监控载具的健康状况,有效载荷输送	飞机系统监控,监控数据链路状况
管理数据	分析传感器数据,监控传感器活动,正面靶标标识跟踪目标	有效载荷数据管理,数据处理,目标探测	传感器操作,外观检查和战术方向	有效载荷指挥和控制有效载荷成果的利用

续表

任务	Nehme 等[8]	Ashdown 等[9]	Peschel 和 Murphy[6]	Ramos[10]
管理任务	分配和调度资源,路径规划监督	资产的任务分派,路径规划,传感器覆盖规划	战略、规划和协调	任务规划和重新规划

2.2.1 可重新配置显示器

无人机系统界面的设计从现代飞行舱使用的多功能显示器概念演变而来。现代主飞行显示器、导航显示器、飞行管理系统和机组警报系统等单元可在许多地面控制显示器中看到。此外,还引入了其他功能,允许对数据链路质量进行监控,并支持任务规划和传感器管理。可重新配置显示器允许在不同的显示模式之间切换,并能实现多种功能,在显示器的使用方面具有更高的灵活性和效率。

2.2.2 感知与规避

SAA 是无人机系统在非隔离空域与载人飞机并肩作战的基本要求。本章引用了多个概念(如 JADEM[11]、MuSICA[12] 和 ACAS Xu[13])作为现有飞机防撞系统的扩展。虽然这些概念提供了在不久的将来实现自主自我分离的途径,但从人因工程的角度来看,人机界面设计应考虑到人为干预的需要(如在系统性能下降或出现异常情况时),提供适当的警告或警示,以及实现及时消除冲突的必要信息。

2.2.3 任务规划和管理

任务规划和管理考虑了实现任务目标所需的不同变量。通常,任务规划包含关于路线、目标和有效载荷以及通信和数据链路质量的信息。支持飞行前任务规划和管理的地面控制功能如下。

(1)优化任务计划的能力。即能够根据不同的任务目标(如最小化燃料、最大化传感器覆盖范围等),基于来自各种来源的输入(如空域和地形数据库,以及天气和交通预测信息)进行优化。

(2)进行任务风险分析的能力。模拟因迫降而对人员和财产造成的潜在损害,以及导航或通信性能的任何损失或退化对整个任务的影响。

(3)应急预案的制定。

(4)执行飞行前预演和模拟。

2.2.4 多平台协作

多无人机协作和控制允许一名操控人员控制多架飞机,要求飞行员的角色向系统管理和任务协调方向有更大的转变,常规任务委托给自动化系统。本章引用了不同的多平台协作范式如下。

(1)蜂群。多个具备自我组织能力的自主平台作为同构实体接受指挥。

(2)对异构载具采用不同管理策略。通常是指具有不同性能限制,配备不同有效载荷,并且具有不同目标的载具。

(3)多操控员、多平台管理策略。操控员可能分布在广阔的地理范围内,协调任务和行动。

表 2.2 对这些操作概念做了进一步阐述。与单一飞行员操作相比,多名飞行员操作要求不同实体之间的更多互动,反过来也反映了复杂性的增加。表 2.2 中描述了不同的操作概念,多平台操作意味着要求无人机平台拥有更高的自动化程度。单平台操作的界面是为了支持人工驾驶任务而设计的,而为多平台操作设计的界面则需要协助遥控操控员监督和管理多个无人机平台,并从 ATC、ATM 或无人机系统交通管理(UTM)显示器的元素中获得灵感,其中操作人员承担协调者的角色[14]。

表 2.2　无人机系统操作概念和人机界面

	单人遥控操控员	多人遥控操控员
单人平台操作	操作概念(1) √手动操作自动化程度低的平台(如低 SWAP-C 微型和极小型无人机系统) √可与单一飞行员通用航空飞机的飞行舱进行对比 操作概念(2) √管理自动化程度相对较高的平台(如 MALE、HALE) √人机界面包括多个显示器,允许地面飞行员监控不同的子系统 √人机界面可以是多功能的,提供一定程度的适应性 √专用显示器可用于执行不同的功能 √可与单一飞行员军用驾驶舱进行对比	操作概念 √控制自动化程度较低且更复杂的无人机 √每个操控员承担不同的功能角色,人机界面根据不同的操控员角色定制 √可与多机组飞行舱进行对比
多人平台操作	操作概念(1) √蜂群,其中多个平台作为单一实体或多个同构实体接受指挥 √可采用几种不同方式来改变蜂群行为[15]: ○改变特定的蜂群特征、策略或算法(如通过行动手册) ○如果需要更高的精度,可直接控制蜂群头机 操作概念(2) √管理多个平台,其中每个平台都被视为一个独立的实体 √平台可以是同构的,也可以是异构的 √可以与空中交通管制站进行对比	操作概念 √协调自主无人机平台 √任务可能在不同操控员之间动态地重新分配 引入以网络为中心的操作元素,要求共享信息和协作决策

2.3　地面控制站人机界面元素

系统设计中的人因工程包括根据系统要求设计合适的人机界面。界面设计涉及定义呈现给操控员的信息内容,以及规定信息的呈现格式,而人机界面的设计则需要定义系统在标称和非标称条件下响应用户和环境输入的行为。良好的界面设计使人机界面能够让操控员有效、高效和持续地完成规定任务,只需进行最少的使用培训,且易于维护。界面设计可以参考物理设计单元(如工作站环境)或数字单元(如专用显示器、警报或控制设备)。

工作站:用户操作环境设计是控制台和工作面在工作站中的位置使操控员能够有效和舒适地执行任务。设计元素包括考虑用户人体测量、环境压力以及安全性。

显示器/警报:旨在以高效和直观的方式向操控员呈现视觉、听觉或触觉信息的设备。呈现的信息可以采取可读消息、图形、图表、指标和符号或话音/声音/动作警报的形式。设计元素可划分为物理特征元素(如显示器尺寸、分辨率、位置等);信息特征元素(如显示格式、警报条件、信息优先级等)和信息内容元素(如准确性、更新率、杂波等)。表2.3列出了这三种元素关键的设计考虑因素。

表2.3　显示器和警报的设计方面和关键考虑因素[16]

设计方面	关键考虑因素
物理特征	✓显示器参数(如屏幕分辨率、亮度、调光和对比度)应可调,以便在操作条件下具有可读性。 ✓显示器应足够大,以容纳可视格式的所有信息。 ✓屏幕或信息刷新率应足够高,以防止闪烁或抖动影响。 ✓显示器的布局应考虑到任何可能影响用户可见度的障碍物。 ✓显示器应按逻辑分组。 ✓视觉/听觉警报应对操控员足够可见/可闻。
信息特征	✓信息应在标称条件下一致地排序。 ✓主视场应包含关键或常用的显示,而次视场(需要用户转头)可包含辅助和咨询信息。 ✓弹出的文本或窗口不应干扰显示器的使用。然而,时间紧迫的警告应出现在操控员的主视场内。 ✓整个飞行舱内的信息显示原理(颜色编码、符号、格式、位置等)应保持一致,以促进熟悉度,并最大限度地减少用户工作量和错误。 ✓信息元素应是明确的,当多个来源提供相同的信息时,允许用户识别信息元素的来源。 ✓相似的信息元素应按照逻辑方式进行分组和排列,以便用户能够轻松识别所需的信息。 ✓应根据紧急程度和关键程度对警报进行优先排序,并且在触发时为用户留出足够时间做出反应。 ✓应避免频繁的误报或漏检,因为它们会增加工作量并降低用户对报警系统的信任

续表

设计方面	关键考虑因素
信息内容	✓设计界面时,应将信息规范(来自需求捕获)用作指南。这可能包括完整性、精度和刷新率要求。 ✓显示/通告信息应有足够的延时性,并且足够详细,以允许用户完成当前任务。 ✓小心、警告或故障标志应显示在其所指或替代的信息位置。 ✓符号的含义应足够清楚,以避免任何误解。 ✓图形元素应以适当的杂波水平呈现,以防止用户混淆,并执行当前任务。 ✓一致的设计理念应考虑条件、呈现、紧迫性和优先顺序

控制:用于提供输入和控制系统的输入、设备和工作面。可能包括按钮、旋钮、杠杆和开关,以及键盘、鼠标、操纵杆、触摸板或声控系统。控制界面的设计方面如表2.4所列,包括控制的物理特征、位置、功能和用户交互的设计。

表2.4 控制界面的设计方面和关键考虑因素(汇总摘自文献[16])

设计方面	关键考虑因素
物理特征	✓应设计为防止堵塞、擦伤、任何形式的干扰或严重磨损。 ✓控制设备的物理外观应通过形式、颜色、位置和标签区分开来(如按钮应与旋钮区分开来)
位置	✓控制设备应按逻辑一致分组和排列。 ✓常用的和/或关键的控制设备应易于访问
功能	✓应尽量减少启动常用控制设备所需的动作数量。 ✓提供多种功能的控制设备应指示当前功能或允许用户轻松查询当前功能。多功能控制设备的使用不应导致工作量、错误率、速度和准确度达到不可接受的水平。应避免隐藏功能或过于复杂的模式。 ✓当需要多个控制设备启动一项功能时,应指定不同的控制设备并贴上清晰的标签,并且在使用过程中不应导致混淆或任何误操作
用户交互	✓应以适合当前任务的分辨率、延时和优先级向用户提供任何必要的信息。 ✓应允许用户根据当前任务以适当的速度、精度和准确度提供输入。 ✓数据输入应允许用户很容易从输入错误中回收错误。 ✓用户应能够超越自动化功能的控制权限。 ✓应提供适当的反馈,通知用户控制模式的任何变化(如在嘈杂的条件下,听觉通知应伴有其他形式的反馈)。 ✓在多机组情况下,反馈能使所有机组成员了解控制的变化。 ✓反馈不应导致误报或漏检。 ✓不得由于界面内或与其他界面之间的任何潜在不一致而导致操控员混淆

设计人机界面时,操控员和自动化系统之间的任务和功能分配是明确的,以及系统性能和它对用户认知过程的影响也是明确的。这些认知过程包括用户的工作负载、态势感知和警惕性。人机界面需要在人类用户和自动化系统共享控制和决策权的情况下精心设计。交互必须确保任务可以在指定的自动化水平上

高效、安全地完成。通过考虑系统或用户的状态,可以使自动化具有适应性,如在系统性能下降期间降低自动化水平并允许用户干预,或者在工作负载或任务复杂性增加时提高自动化水平以提供额外帮助。表2.5中给出的10个自动化水平以Endsley和Kaber的分类法[17]为依据,包括以下四个功能。

表2.5 自动化水平[17]

自动化水平	功能			
	监控	生成	选择	实施
1. 手动控制	人	人	人	人
2. 行动支持	人/计算机	人	人	人/计算机
3. 批处理	人/计算机	人	人	计算机
4. 共享控制	人/计算机	人/计算机	人	人/计算机
5. 决策支持	人/计算机	人/计算机	人	计算机
6. 混合决策	人/计算机	人/计算机	人/计算机	计算机
7. 刚性体系	人/计算机	计算机	人	计算机
8. 自动化决策	人/计算机	人/计算机	计算机	计算机
9. 监督控制	人/计算机	计算机	计算机	计算机
10. 全自动化	计算机	计算机	计算机	计算机

(1)通过视觉、听觉或其他方式监控系统状态。
(2)形成实现目标的方案或策略。
(3)选择策略。
(4)实施选定的策略。

2.4 人因计划

人因工程(HFE)是指将关于人的能力和局限性的知识应用于系统和/或设备的设计/开发,以最低的成本、人力、技能和培训需求实现高效、有效和安全的系统性能。人因工程保证了工作环境、设施和设备以及系统和用户任务的设计与负责支持、维护和操作这些设备的人员的身体、感知和认知属性一致。制定人因工程计划旨在支持特定硬件、软件、系统或工作环境的设计、测试和集成,以满足人机性能需求,可通过以下方式实现。

(1)定义完成操作或任务目标所需的性能需求。
(2)在分析所需任务和功能分配的基础上建立工作流程。
(3)设计合适的界面,包括硬件、软件和系统单元以及工作环境,以使人类用户能够成功地执行他们的任务。

(4)识别可能损害整个人机性能的安全性、可靠性、效率或有效性的关键任务和设计要素。

(5)通过测试和评估来评价用户和系统性能,并获得对设计产品可用性的反馈。评估用于系统界面或工作程序的后续再设计。

(6)评估系统生命周期成本,包括运行、维护、支持和培训成本。

FAA/HF-STD-004 提供了有关人因工程通用需求和评估程序的更多细节[18]。

如图 2.4 所示,人因工程计划包括要求的三个主要阶段:定义、设计和评估。这三个阶段通常在多个周期内迭代,逐步迭代为需求、过程和系统设计提供更多细节并进行完善。应对定义、设计和评估活动进行规划,以便同步进行,同时还具备连续评估的灵活性,以影响未来阶段。具体而言,人因工程计划包括以下活动。

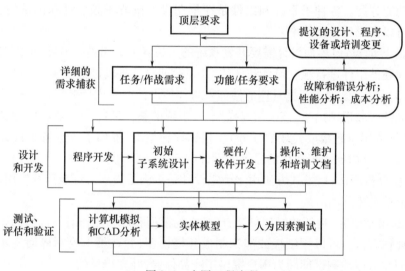

图 2.4 人因工程流程

(1)需求定义:需求为确定具体设计元素提供了一个框架,也支持根据具体目标对这些设计元素进行评估,从而推动设计和评估过程。初期需求可基于先前设计阶段的评估进行完善。

① 顶层需求开发:顶层需求源自客户,它规定了产品用途(正在设计的硬件、软件或系统)。顶层需求构成了设计和开发阶段的初始技术基线,因此应该充分获取相应的细节。

② 定义任务和操作需求:这些需求定义了任务环境、目标、操作限制条件以及性能和人力需求。操作场景是评估系统性能和成功的典型案例——可以根据设计和评估需求对场景进行修改和完善。

③ 获取功能和任务需求：这些需求定义了完成拟定任务所需的功能和任务。详细说明了任务及其执行条件。此类需求与任务和操作需求类似，在设计过程中不断完善。

（2）设计：产品设计包括设计和开发程序、软件、系统和设备，符合在需求定义阶段建立的规范。初始概念设计为后续改进提供了框架。根据需求，初始设计应明确信息流、任务次序、系统布局以及任务和功能的分配。随后，可用更具体的需求和设计目标来评估产品设计，从而可进行更详细的改进。为了确保有效的开发周期，人因原则应该贯穿整个设计过程，而不是仅用于详细设计的最后阶段。

① 程序开发：包括规定人员配置、操作、维护和培训的程序和指南，以及相应文档。

② 系统设计：包括定义系统架构、信息流和系统/子系统各种模块和组件之间的功能分配。各种子系统和组件是在早期设计阶段单独设计和测试的，方便以后可以集成到主系统中。

③ 界面设计：界面设计需要开发系统的人机界面2组件。另外，可以对各个软件和硬件组件进行独立设计和评估，并在后期设计阶段进行集成。

（3）评估：该过程根据具体的性能需求评估设计在实现任务目标方面的有效性。关键区域识别出来，该区域里操控员/系统的整体性能和安全性可能受到影响（如导致用户超载或欠载的任务）。根据评估情况，提出了改进特定设计元素的建议，并反馈到未来迭代的需求和设计中。

① 分析评估：启发式方法用于确定设计是否满足需求。可能涉及使用框图、检查清单和矩阵。

② 模型研究：在评估过程早期，产品实物模型和原型用作对设计概念相对快速的评估方法。可以用纸板或泡沫制成环境模型，并在纸上绘制界面显示。这些模型在可用性测试期间为用户提供了产品体验，并支持评估。

③ 计算机辅助评估：采用计算机辅助设计（CAD）工具，通过模拟虚拟工作环境中虚拟操控员来评估人类环境因素。在更成熟的评估阶段，仿真器用于评估虚拟工作环境中真实操控员的表现。软件验证和确认以及系统级测试工具也用于计算机辅助评估。

④ 人工测试和评估：人在回路评估是为了评估系统在典型任务环境中的性能。评估分阶段进行，通常是从测试单个组件到评估集成系统再到全面评估，范围逐渐扩大。主用户（工作负载/可用性）评级来评估人为性能方面，如工作负载、态势感知和可用性。

⑤ 权衡研究：用于在特定设计特征之间做出决定，通常根据实施和生命周期成本、满足任务目标的有效性、可用性以及操控员安全和风险等指标进行评估。

2.4.1 需求定义、获取和改进

需求在整个软件开发过程中得到改进、修改和应用。因此,在整个开发生命周期中,对需求的正确管理、记录和维护至关重要。表 2.6 示出了服务于不同功能的各种类型的需求。

需求应满足以下属性:

具体:需求必须具体,因为系统设计或性能只有一个方面必须得到解决。必须根据需要(需要什么,为什么需要,需要做到何种程度)来明确表达需求,而不是规定如何达到需求。

表 2.6 不同类型的需求

需求类型	需求说明
客户需求	关于客户对系统期望的事实陈述或假设,涉及任务目标、运行环境和约束条件,以及有效性和适用性衡量标准
功能需求	必须完成的必要任务、行动或活动,随后在执行功能分析时用作顶层功能
性能需求	执行使命、功能或任务所必须达到的程度;通常以数量、质量、覆盖面、及时性或准备程度来衡量
设计需求	在技术数据包和技术手册中表达的需求,规定了产品构建/编码/购买以及使用所涉及过程中的需要
衍生需求	更高层次需求中隐含的或从中产生的需求
划分产生的需求	一个较高层次的需求被划分或分为多个较低层次需求时所产生的需求

(1)可测量:需求必须有可测量的性能指标,并且必须以客观且最好以量化方式进行验证。

(2)可实现:需求必须反映一种需求或目标,在这种需求或目标下,在技术上找一个成本相对可承受的解决方案。

(3)相关性:需求必须具备相应的详细信息,用于指定的系统层次。它不应该过于详细,而限制了当前设计水平的解决方案(如系统级规范不包含详细的组件需求)。

(4)可追踪:较低层次的需求必须源自较高层次的需求,并支持较高层次的需求。需要评估无法追踪到较高层次的需求,以判断是否有必要纳入。

(5)完整性:需求必须能够清楚表明客户的需要(如操作概念、使用环境和限制条件、任务简介和维护需求)。

(6)一致性:需求不能存在冲突,必须解决冲突以确保一致性。

需求工程化过程如图 2.5 所示,包括需求收集、分析、规格、验证和记录以及建模的迭代循环。在收集阶段,通过访谈、焦点小组或团队会议来获取需求。对需求进行分析,确定领域约束,并纠正需求中的任何缺口和模糊性,以确保一致性。然后可以在规范阶段对它们进行分类、确定优先顺序和正式表达(如作为图形或数学模型)。验证阶段审查规范以确保规范一致、正确、可理解、实际、可测试和可追踪。当规范达到要求时,需求和模型被适当记录下来。

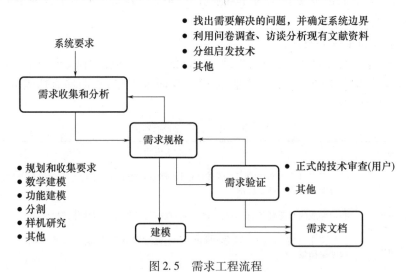

图 2.5　需求工程流程

2.4.2　任务分析

为了支持以用户为中心的系统和设备的设计和开发,人机界面围绕用户的需要和需求进行设计,任务和功能分析用于将顶层需求转化为工作程序和具体的设计标准。功能是实现特定目标所必需的离散过程,任务是实现特定功能所需的连续步骤;它们通过使用设施、设备、软件、人员或其组合来执行。通常,首先通过访谈、调查或对现有标准/文献的审查来收集关于功能和任务的信息,从而进行分析。在分配过程中,根据逻辑顺序划分功能,以最大限度地减少用户工作负载和控制界面。设计过程生成一个功能架构,在该架构中,按照系统组成部分的功能对系统进行说明。部分任务和功能分析工具将在下面讨论。

2.4.3　分层任务分析

分层任务分析以分层的方式采用框图组织任务和功能,框图体现目标和任务之间自上而下的关系。顶层目标分解为低层子目标,低层子目标本身又可以分解为低层任务和子任务。因此,一个复杂的任务可以用一个简单行动的层次来表示,当任务按顺序进行时,就可以实现上级目标。每个任务的细分程度(或子任务的数量)取决于几个因素,如任务的复杂性或关键程度、人为错误的风险或特定任务的复杂性。分层任务分析允许以不同的详细程度来说明任务的不同方面,并作为组织和表示活动的有效方法。允许对任务的层次结构进行初步分析,以确定人类用户的工作负载,从而支持系统功能的后续分配。

2.4.4 认知任务分析

为了识别和诊断具有更复杂人机界面的任务的特定错误和故障模式,可以执行认知任务分析来了解人类用户的基本心理过程。认知任务分析从描述任务、子任务和用户控制动作的活动列表(如从总体任务分析[19]中获得的活动)开始。然后,通过分析特定场景中的认知线索、信息流需求和决策策略,对列表进行扩展和更新,以模拟操控员的心理过程。如 Salmon 等[20]以及 McIlroy 和 Stanton[21]所述,一个五阶段的分析过程用于分析工作领域、控制任务、策略、社会组织与协作,以及工作人员的能力。与分层任务分析类似,每阶段都采用不同方法进行信息获取和表述。由于说明充分,认知工作分析最终可以作为一种工具来形成正式的程序指南和需求说明。

决策阶梯是认知任务分析[22]的结果之一,说明了用户在执行任务时决策过程中每个阶段的信息处理活动。这组路径代表了通过决策阶梯的决策策略,操控员可以使用,并且可以包含不同级别的情况分析/诊断和行动计划。例如,操控员工作可以基于启发式方法、预先定义的操作程序或由于时间限制,选择绕过某些过程的路径。

认知任务分析提供操控员所需的认知过程信息,确定可用的认知策略和需求,并最终用于推定相关人因需求的系统级定义。这些需求包括信息和界面需求、系统性能需求以及培训先决条件。

2.4.5 关键任务分析

关键任务分析(CTA)用于评估安全和关键业务任务中涉及的人为因素。该过程类似于认知任务分析,包括从关键任务中获取数据,使用表格和/或图表来表示任务,以及识别所有操作场景(包括降级和紧急情况)中影响性能的故障模式和因素。CTA 可用于系统设计,以尽可能降低风险并提高性能。FAA – HF – 004A[23]规定了 CTA 报告(CTAR)的格式和内容需求,并包含表 2.7 中所述项目。

表 2.7 CTAR 的内容需求列表

信息类别	信息说明
任务层次关系	✓完成任务所需的行动。 ✓人为错误的概率和严重性。 ✓错误恢复的可能性

续表

信息类别	信息说明
信息流	✓操控员/维护人员要求并可获得的信息。 ✓反馈，告知操控员/维护人员所采取措施的充分性。 ✓所需的通信，包括通信类型。 ✓操控员/维护人员交互
认知过程	✓决策评估过程。 ✓评估后做出的决定。 ✓所需人员数量；人员的专长/经验。 ✓人员的行动限制
实际执行过程	✓为完成任务采取行动。 ✓完成任务的可用时间。 ✓采取行动所需的身体移动。 ✓动作的频率和偏差。 ✓所需人员数量；人员的专长/经验。 ✓人员的行动限制
工作环境	✓需要/可用于采取行动的工作区范围。 ✓工作环境的位置和条件。 ✓所需工具和设备。 ✓所需工作辅助、培训或参考。 ✓所含危险。 ✓硬件/软件的操作限制

2.4.6 操作顺序图

操作顺序图是说明特定活动工作流程的流程图。任务和功能按照执行的顺序依次排列，在多个操控员和自动化功能之间划分为图中的不同通道。操控员和系统之间的信息流由代表不同功能或动作的独特符号表示。操作顺序图提供关于决策行为、通信链接、相互关系、任务频率、任务时间和工作负载的信息，为人因工程师、系统设计师和软件工程师提供特定活动所需的任务和功能的通用参考。基于图中确定的功能和信息流，可以设计和开发相应的子系统、接口和交互。

2.4.7 系统设计和开发

软件设计将需求转化为四类设计细节或活动，提供一个软件设计规范，用于说明数据、架构、接口和组件。

（1）数据设计是将信息模型转换成数据结构，允许软件有效地检索和存储数据，并将程序组织成模块，从而降低软件的整体复杂性。规定了数据结构的类型、数据结构之间的链接以及数据的完整性规则。

（2）架构设计包括确定软件中的模块并定义这些模块之间的关系。架构设计细分为三类：概念架构，即确定不同的软件部件，并将任务分配给每个逻辑架构；逻辑架构，侧重部件接口、交互、连接机制和协议的设计；执行架构，侧重于软件组件在运行时的通信、协调和资源分配。

（3）接口设计包括设计不同系统组件之间以及程序和终端用户之间的接口。它说明了软件如何同与之交互操作的系统进行内部通信，以及如何通过数据和控制流元素与用户进行通信。

（4）构件级设计将软件架构的结构元素转换成软件组件的程序说明。

航空电子和 ATM 系统开发中使用的典型语言有 C 语言、C++和 Ada。C 语言是一种面向过程的程序语言；C++支持面向对象的编程范式；Ada 也支持面向对象编程，并在各种安全关键的应用程序中采用。Ada 拥有许多特性，允许开发可靠的程序，也增强了代码的安全性和可维护性。近年来，嵌入式航空电子和 ATM 系统的开发也采用了 Java 和 Python 语言。

企业架构描述了一个组织的结构和运作，使该组织能够在战略上团结一致地运作。架构框架为创建、使用和分析架构提供了标准基础。开放组架构框架（TOGAF）是一个广泛使用的框架，用于设计、规划、实施和维护企业信息技术（IT）架构。TOGAF 支持四个相关领域的企业架构设计。这些领域包括满足用户、规划者和业务管理需求的业务架构；满足系统和软件工程师需求的应用架构；满足数据库设计人员、管理员和系统工程师需求的数据架构；以及满足硬件/软件购买者、操控员和管理员需求的技术架构。

其他框架包括英国国防部体系结构框架（MODAF）、美国国防部体系结构框架（DODAF）和北约体系结构框架（NAF）。这些架构框架在国防背景下用于开发管理复杂系统级系统的架构，从而确保网络基础设施内的互操作性，并实现不同利益相关群体之间的有效决策。

基于模型的系统/软件工程是贯穿软件生命周期不同阶段的模型应用。基于模型的系统/软件工程利用行业标准建模语言，如统一建模语言（UML）、系统建模语言（SysML）、DODAF 或 MODAF 来支持软件和系统的规范、分析、设计、验证与确认（V&V）。IBM 的 Rational Rhapsody 等集成开发环境（IDE）为团队提供了一个建模环境，以协作开发复杂的嵌入式应用程序。

2.4.8 设计评估

在评估阶段，根据设计需求对系统输出进行评估，以确定产品是否能符合用户的期望运行，评价指标包含可行性、可用性、完整性、可靠性和对操控员绩效的影响等指标。评估阶段还确定了操作过程中的潜在安全隐患、故障模式或意外交互。可以在不同真实水平下对独立部件或更大范围集成系统的一部分进行评估。之后,评估结果可用于改进需求,解决设计中的任何潜在风险和问题,并对已识别风险和问题的设计领域提出修改建议。根据设计周期中的成熟度以及时间和成本需求,采用不同的评估技术。

工程分析和评估提供了对设计输出、任务工作流程、功能以及设计的物理和认知人机工程学方面的初步评估。常见技术包括功能和识别测试、CTA、人为错误、可靠性分析和风险评估(HRA)或故障模式和影响分析(FMEA)。根据设计需求和/或人为因素检查表对设计方面进行测量和评估。然后可以根据获得的反馈对更改提出建议。确定了关键任务和功能,以及进一步设计的详细要求。关键评估指标已确定,可在后续评估中进行测量,以确定是否满足性能需求。初步评估相对灵活,因为它可以适应不同程度的详细情况,并且对成本和时间的需求相对较低。应在设计过程的早期采用初步评估,以确保良好的实践方法贯穿始终,且初步评估是作为更详细的评估方法的初始规划活动。由于评估是在概念层面上进行的,因此会对产品、操作环境和用户做出假设,且这些假设必须在后期设计阶段进行完善。

部分任务评估包括在典型条件下对设计的有限评估,终端用户通常会提供反馈。常见方法包括访谈、演练分析和可用性测试。界面、硬件和模拟系统的组合用于评估人类用户在执行特定任务集时的表现。评估的人为因素可能包括界面可见性、与不同自动化模式的交互、所呈现信息的可用性/准确性、操控员态势感知和工作负载,以及任务性能指标,如错误率、准确性和执行时间。部分任务评估对独立部件或任务进行评估,允许并行评估或故障/错误分离。此外,由于评估通常建立在先前的任务分析之上,任务评估程序可以随着产品的成熟而逐步完善,从而实现设计过程中的迭代改进。与全系统评估相比,独立任务的评估更具包容性,且相对更高效。然而,部分任务评估可能无法捕捉更大、更复杂系统的性能,并且不同任务和系统的交互可能会产生意外漏洞。此外,如果需要评估一个部分工作的系统,只能在开发过程的更成熟阶段进行。

模拟和实物模型允许在试验室环境的典型条件下评估设计输出。评估的真实性取决于产品集成到主机平台的程度、场景设计的详细程度、模拟环境的复杂性以及操控员的参与程度。由于模拟的逼真度可以变化,可用在设计过程的诸多阶段,并且也提供了很大的灵活性,由于它们允许对快速原型到详细模型和系统进行评估。独立任务评估中通常无法捕捉到的复杂依赖关系和交互可能会在实际模拟中出现。此外,在受控的环境中进行评估,允许更全面的数据收集、可重复的测试和相对于现场测试而言更低的风险。然而,复杂的实体模型和模拟开发成本高、耗时长,而且维护模拟器/试验室的成本也很高。

现场测试提供了对产品在外部(现场)环境中高度真实的操作场景的评估。

评估通常在开发过程的成熟阶段进行,此阶段对最终设计的运行有效性/适用性进行评估,或者在运行阶段对产品的性能在其整个生命周期内持续监控。现场测试对产品在操作条件下的稳健性/适用性进行高度真实的评估,但相对于试验室评估而言,通常更耗时且成本更高。此外,在外部环境中的测试意味着试验控制低,安全性和风险成为更关键的考虑因素。

2.4.9 验证与确认

软件的验证与确认包括:① 产品验证,以保证其预期功能没有错误或意外情况,并且足以对响应异常输入和条件做出适当的反应;② 产品批准,以保证客户对高层次需求满意,前提是这些需求已被适当地转化为低层次需求、已开发的软件架构和源代码。验证与确认根据系统预期运行环境中的系统需求对系统功能进行评估,并用于发现任何系统缺陷。RTCA DO-178 传统上是进行验证与确认的参照点,且截至撰写本章时,DO-178B 已经扩展到包括基于航空电子的软件和 DO-178C 和 DO-278A 中地面和空中交通管理的软件。关于 RTCA 文件的进一步讨论见参考文献[24]。

在验证与确认中使用了静态和动态两种方法。在静态验证中,通过文件和代码分析工具检查软件以检测错误。软件检查用于验证软件是否符合给定规范,但不能验证非功能特性,如性能和可用性。静态分析器等工具可以用来补充检查,并通过解析程序文本来发现潜在的错误情况。静态分析的阶段如下。

(1)控制流分析:检查具有多个出口或入口点的循环,寻找执行不到的代码等。

(2)数据使用分析:检测未初始化的变量、两次写入但没有中间赋值的变量、已声明但从未使用的变量等。

(3)接口分析:检查例程和程序声明的一致性。

(4)信息流分析:确定输出变量的依赖性。不检测异常本身,但突出代码检查或审查的信息。

(5)路径分析:确定程序中的路径,并列出在该路径中执行的语言。

动态验证与确认通过软件测试进行,测试数据用于运行系统,以便观察运行性能。进行不同类型的测试,如发现系统缺陷的缺陷测试,以及确定用户输入频率和估算程序可靠性的统计测试。测试可以显示错误的存在,但不能显示错误的不存在,如果发现一个或多个错误,则认为测试成功。如果发现缺陷或错误,程序将进行调整,以定位并修复错误。

表 2.8 概述了飞机软件和系统开发周期中所用的主要系统安全评估部分。这些组成部分包括功能危险评估(FHA)、初步系统安全评估(PSSA)、系统安全评估(SSA)和共同原因分析(CCA),并支持系统安全需求的系统定义、评估和保证。

表 2.8　系统安全评估的组成部分[25]

主要系统安全评估组成部分	说明
FHA	FHA 在开发周期开始时进行,以确定故障情况和与每个功能相关的严重程度分类。FHA 允许为特定的故障情况和随后的定量概率需求分配安全目标
PSSA	根据 FHA 确定的故障情况,PSSA 对系统架构进行检查,以制定安全需求,确保达到安全目标。PSSA 涉及对从系统功能架构中衍生出的功能进行故障树分析(FTA),并计算故障概率
SSA	SSA 使用类似于 PSSA 活动的方法,对系统设计进行更全面的评估。然而,鉴于进行 PSSA 是为了确定系统安全需求,SSA 则验证拟定设计是否满足 FHA 和 PSSA 分析中规定的需求。SSA 涉及可靠性数据的计算和达到特定安全目标所需的冗余水平的测定
CCA	CCA 评估集成系统或功能的故障模式,以确保单个系统的故障模式相互独立,或者与任何现有依赖性相关的风险可接受。CCA 包括三种类型的分析:区域安全分析(ZSA)、特殊风险分析(PRA)和共模分析(CMA)

2.5　未来的工作

随着工业 4.0 的发展,必须考虑新的发展。研究表明,工业 5.0 将推动① 人机合作,② 协同无人机,③ 人工智能和机器学习的进一步应用。

感官替代竞技舞台将使得可为有感官障碍的个人打造美观、小巧、成本低且可广泛使用的设备。此外,研究者控制感觉替代过程时,可以引入新的工具来探索知觉、身体和大脑功能与复杂认知过程之间的关系和相关性。

研究表明,依靠感官替代来重建缺失的感官功能的方法是值得信赖的。在 UAV-CPS 人机界面设计中,需要处理如下三个主要方面。

(1)设计稳健且价格相对低廉的技术部署,让更多有感官缺失的人员能够使用。

(2)扩展人类的感知能力,如在不干扰正常视力的情况下实现夜视。

(3)加速采用人体非侵入式安全研究,以提高大脑除认知过程外的可塑性认知。

2.6　小结

人机界面是无人机系统开发中关键但经常被忽视的方面。适当规划的人机

第 2 章 无人机系统人为因素和人机界面设计

界面提高了态势感知能力,减少了地面飞行员手中的任务,从而提高了总体任务完成率。人因工程经常使用方法程序和知识来协助人机界面设计。迭代阶段有三个方面:① 先决条件确定和规范;② 框架设计;③ 评估。各种方法都可以为人机界面元素的人为、人机工程学和操作工程方面贡献力量。考虑到各种无人机系统的用途和任务广泛,以用户为中心的开发实践有助于明确定义开发过程,其中人机界面是根据对人类用户和操作框架的需求分析而设计的。无人机系统的日常作用包括任务组织、感知、证据探索和 SAA,可以扩展到多平台协调和协作管理。对人为因素和支持这些功能的相关 HMI 组件的研究是需要大量研究的关键领域,并将通过新兴技术和模型原型来推进。

参考文献

[1] F. Kendoul,"Towards a Unified Framework for UAS Autonomy and Technology Readiness Assessment (ATRA)," in *Autonomous Control Systems and Vehicles. Intelligent Systems*, *Control and Automation*: *Science and Engineering*, vol. 65, K. Nonami, M. Kartidjo, K. J. Yoon, and A. Budiyono, Eds., Tokyo: Springer, 2013, pp. 55 – 71.

[2] K. W. Williams,"Human Factors Implications of Unmanned Aircraft Accidents: Flight Control Problems," DTIC Document 2006.

[3] G. L. Calhoun, M. A. Goodrich, J. R. Dougherty, and J. A. Adams,"Human Autonomy Collaboration and Coordination Toward Multi – RPA Missions," in *Remotely Piloted Aircraft Systems*: *A Human Systems Integration Perspective*, N. J. Cooke, L. Rowe, W. Bennett Jr, and D. Q. Joralmon, Eds., UK: John Wiley & Sons Ltd, 2017.

[4] N. J. Cooke and H. K. Pedersen,"Chapter 18. Unmanned Aerial Vehicles," in *Handbook of Aviation Human Factors*, J. A. Wise, V. D. Hopkin, and D. J. Garland, Eds., Boca Raton: Taylor and Francis Group, 2009.

[5] R. Hopcroft, E. Burchat, and J. Vince,"Unmanned Aerial Vehicles for Maritime Patrol: Human Factors Issues," Defence Science and Technology Organisation, Edinburgh (Australia) Air Operations Div 2006.

[6] J. M. Peschel and R. R. Murphy,"Human Interfaces in Micro and Small Unmanned Aerial Systems," in *Handbook of Unmanned Aerial Vehicles*, Springer, 2015, pp. 2389 – 2403.

[7] T. Porat, T. Oron – Gilad, M. Rottem – Hovev, and J. Silbiger,"Supervising and Controlling Unmanned Systems: A Multi – Phase Study with Subject Matter Experts," *Frontiers in Psychology*, vol. 7(568), 2016, pp. 1 – 17.

[8] C. E. Nehme, J. W. Crandall, and M. Cummings,"An Operator Function Taxonomy for Unmanned Aerial Vehicle Missions," in *12th International Command and Control Research and Technology Symposium*, Washington, DC, 2007.

[9] I. Ashdown, H. Blackford, N. Colford, and F. Else,"Common HMI for UxVs: Design Philosophy

and Design Concept," *Human Factors Integration*, *Defence Technology Centre Report*, *BAE Systems*, vol. 17, p. 18, 2010.

[10] F. J. Ramos, "Overview of UAS Control Stations," in *Encyclopedia of Aerospace Engineering*, West Sussex: John Wiley & Sons, Ltd, 2016.

[11] R. C. Rorie and L. Fern, "The Impact of Integrated Maneuver Guidance Information on UAS Pilots Performing the Detect and Avoid Task," in *Proceedings of the Human Factors and Ergonomics Society Annual Meeting*, 2015, pp. 55 – 59.

[12] R. H. Chen, A. Gevorkian, A. Fung, W. - Z. Chen, and V. Raska, "MultiSensor Data Integration for Autonomous Sense and Avoid," in *AIAA Infotech@ Aerospace Technical Conference*, 2011.

[13] G. Manfredi and Y. Jestin, "An Introduction to ACAS Xu and the Challenges Ahead," in *Digital Avionics Systems Conference (DASC)*, 2016 IEEE/AIAA 35th, 2016, pp. 1 – 9.

[14] A. Fisher, R. Clothier, P. MacTavish, and J. Caton, "Next – Generation RPAS Ground Control Systems: Remote Pilot or Air Traffic Controller?," in *17th Australian International Aerospace Congress (AIAC 2017)*, Melbourne, Australia, 2017.

[15] A. Kolling, P. Walker, N. Chakraborty, K. Sycara, and M. Lewis, "Human Interaction with Robot Swarms: A Survey" *IEEE Transactions on Human – Machine Systems*, vol. 46, pp. 9 – 26, 2016.

[16] M. Yeh, J. J. Young, C. Donovan, and S. Gabree, "FAA/TC – 13/44: Human Factors Considerations in the Design and Evaluation of Flight Deck Displays and Controls," Federal Aviation Administration, Washington DC, USA, 2013.

[17] M. R. Endsley and D. Kaber, "Level of Automation Effects on Performance, Situation Awareness and Workload in a Dynamic Control Task," *Ergonomics*, vol. 42, pp. 462 – 492, 1999.

[18] FAA, "HF – STD – 004: Requirements for a Human Factors Program," Federal Aviation Administration HF – STD – 004, 2009.

[19] R. B. Miller, "A Method for Man – Machine Task Analysis," DTIC Document 1953.

[20] P. Salmon, D. Jenkins, N. Stanton, and G. Walker, "Hierarchical Task Analysis vs. Cognitive Work Analysis: Comparison of Theory, Methodology and Contribution to System Design," *Theoretical Issues in Ergonomics Science*, vol. 11, pp. 504 – 531, 2010.

[21] R. C. McIlroy and N. A. Stanton, "Getting Past First Base: Going All the Way with Cognitive-WorkAnalysis," *Applied Ergonomics*, vol. 42, pp. 358 – 370, 2011.

[22] J. Rasmussen, "Outlines of a Hybrid Model of the Process Plant Operator," in *Monitoring Behavior and Supervisory Control*, Springer, 1976, pp. 371 – 383.

[23] FAA, "FAA – HF – 004A: Critical Task Analysis Report," Federal Aviation Administration FAA – HF – 004A, 2009.

[24] S. A. Jacklin, "Certification of Safety – Critical Software Under DO – 178C and DO278A," in *AIAA Infotech@ Aerospace Conference*, GardenGrove, CA, 2012.

[25] SAE, "ARP4761: Guidelines and Methods for Conducting the Safety Assessment Process on Civil Airborne Systems and Equipment," Pennsylvania, USA, S – 18, 1996.

第3章 无人机中的开源软件和开源硬件

近些年来,开放源码工具(OST)的普及程度有了显著的提高。同样,基于开源硬件(OSH)的无人驾驶飞行器(UAV)也是如此。开源软件(OSS)和OSH可以应用于广泛的领域,并且可以对多种技术的改进提高有所帮助。本章以对OSS的介绍作为开头,描述其基本原理,阐述OSS和专用软件(PS)之间的基本区别,OSS对整个无人机领域的好处,以及引导人们选择OSS而不是PS的动机,这对于学术界和研究界来说都是有帮助的。本章还涵盖了一些在无人机集群中用于对无人机的各个方面技术以及遥感(RS)和摄影测量数据的后期处理链提供支持的OSS。仅仅基于OSH和OSS来建立完全自动化和可操作的无人机是完全有可能的。本章介绍了在无人机科技中广泛使用的最新OSS技术,以及无人机各个方面技术所使用的软件,包括基于ARDUPILOT的Autopilot固件、基于MISSION PLANNER的地面站、OPENTX发射机软件、MINIM On-Screen Data(OSD)软件、Open Drone Map摄影测量数据处理套件、网络无人机数据处理套件WebODM。本章介绍了开源软件/硬件、内置功能、特定功能以及平台要求的一些概念和特点,并提供了一个典型的无人机摄影测量工作流程,用于具有飞行计划/执行和OSS数据处理能力的无人机的建设。

3.1 引言

微型飞行器(MAV)是指有5kg的最大起飞载重量(MTOW),大约1h的飞行续航时间和10km左右的操作范围的无人驾驶飞行器(UAV)种类。微型飞行器在摄影测量和遥感(RS)领域有着广泛的应用。飞行器若想在传统摄影测量中有更大潜在应用,就需要有更低成本的替代方案。微型飞行器的构建和软件开发主要有两种思路:开源硬件(OSH)和开源软件(OSS)。业余爱好者和研究人员主要利用OSH和OSS制造微型飞行器。最富有成效的开源项目都有学术根基,并且现在项目的进行要参考各种各样的学术活动。本章重点介绍了开源项目的优缺点。尽管有高质量和高精度的选择,这种由私人公司开发、支持和维护软件和硬件的纯商用微型无人机模式可能会限制无人机的发展,特别是在世界危机之后。

3.2 开源软件

首先,解释开源的概念是最为重要的[1],它指的是可调控和共享的软件或硬件。它的设计是完全公开的,并与更广泛的背景(也称开源方式)相关联。OSS和OSH的项目、产品和计划涵盖了开放交流、合作、透明、快速原型、精英管理和面向集群的改进。开源软件是一种包含源代码的软件,并且任何人都可以查看这种源代码、修改和补充。在计算科学中,源代码包含一组带有注释的代码,它们是用适应人类习惯的编程语言编写的,通常是纯文本[2]。程序源代码被专门设计用来对由编写的源代码主要完成的计算机执行任务进行平滑处理。编译器通常将源代码转换为目标机器(PC、Arduino Board、Android 等)能够理解的二进制代码。大多数应用软件仅可作为可执行文件来传播。如果程序中包含了源代码,那么其对任何想要研究或修改程序的人都会有所帮助。

由于程序员可以向源代码添加功能或修改不总是正常工作的部分,因此任何人都可以改进程序。Mission Planner、Ardupilot、OpenTX、OpenDroneMap、WebODM 就是 MAV 和摄影测量 OSS 的示例。

在闭源的情况下,源代码只有创建它的个人、团队或组织才能维护、控制和修改。在商业或专有软件(PS)的情况下,只有闭源软件的原始作者才能复制、检查和返工该软件。目前,几乎所有的商用微型飞行器都依赖于私有软件。DJI(大疆)创新、Parrot SA、Yuneec 等公司提供商业产品;微型飞行器来运行闭源软件,因而其他程序员无法修改其源代码。

一般来说,OSS 了促进协作和共享,这允许其他人调整源代码并将这些修改集成到他们的项目中。在公用无人机中,一个非常生动的例子是 ArduPilot 公司和组织。目前,Ardupilot 是最先进、功能最全、最值得信赖的开源自动驾驶软件[3,5-6]。多年来,它由合格的工程师、计算机科学家和公司开发。自动驾驶软件可以调节任何机器人车辆(如常规飞机、多旋翼、直升机、船只和潜艇)。开源代码的概念意味着它总是处于技术的最前沿并快速发展。由于有许多外围供应商和接口存在,用户得以受益于由传感器、配套计算机以及通信系统组成的全方位生态系统。总之,可以评估开源代码以确保遵守其安全性和机密性。如参考文献[1]所述,由于某些原因,人们更喜欢 OSS 而非 PS,包括以下方面。

（1）控制。选择 OSS 有利于对软件实现更多的控制。可以测试代码以确保如果 UAV 没有做任何事情，那么控件应该保持休眠状态。

（2）训练。开源代码可以帮助人们成为更好的程序员。随着开发的进行，学生、调查人员或其他任何人都可以有效地研究可公开访问的代码以制作更好的软件，并与他人分享他们的工作，以及邀请别人来评论和批评。当在程序的源代码中发现一些错误时，可以与其他人共享，以避免重复相同的错误。

（3）安全。与 PS 相比，由于其令人满意的安全性和稳定性，有些人更喜欢 OSS。由于任何人都可以评估和修改 OSS，因此有人可能会识别到并修复程序原始作者未发现的错误或遗漏。程序员可以比 PS 更快地修正、更新和升级 OSS。

（4）稳定。由于源代码是公开散播的，因此在重要且长期存在的项目中需要 OSS，因为用户可以依赖该软件进行关键活动，并且不必担心如果原始开发人员停止该项目，他们的工具会消失或降级。OSS 也倾向于根据开放标准（图 3.1）来整合功能。

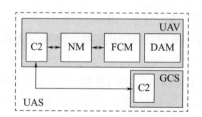

图 3.1　UAS(UAV – CPS)系统基本模块[4]

3.3　开源 UAS

UAV 是名为 UAV – CPS 或 UAS 的网络 – 物理系统（CPS）的一部分，除了包括 GCS 和 UAV 之间的通信和控制（C^2）链路之外，它还包括一架无人机和至少一个地面控制站（GCS）[4]。如果完整的 UAV – CPS 采用了 OSH（图 3.2）和 OSS，则可以称其为开源无人机系统（OSUAS）。这意味着 UAS 中使用的所有主要组件和软件都采用开源思想。

UAV 的关键构建模块是机载 UAV 平台，如导航模块（NM）、飞行控制模块（FCM）和机械伺服系统。根据 UAV 的主要用途和相关任务，其有效载荷可能会有所不同。对于摄影测量和 RS 来说，其有效载荷是数据采集模块（DAM）[5-9]。

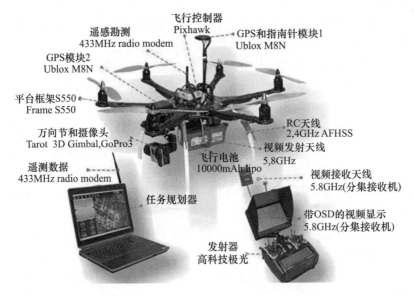

图 3.2 OSH 和基于软件的 UAS[5]（见彩图）

NM 是机载最关键的模块，它是一种可反复向 FCM 提供飞机的位置、速度和高度的无人机。因此，NM 可以向 FCM 提供关键的无人机导航数据。NM 包含用于确定平台位置的导航系统（NS）（通常通过 GNSS 实现）和带有加速度计（运动传感器）和陀螺仪（旋转传感器）的定向系统，从而其可以持续估计惯性测量单元（IMU）平台的朝向、运动方向和速度。自动驾驶仪包含 NM（NS 和 OS）和 FCM，并且它们集成在一个模块中。

FCM 是飞行器的指挥设备，这意味着其要在设置正确的方向和速度的同时将 UAV 引导到选定的位置。FCM 包含两个部分：① 数据分析台，其用来接收来自系统操控员的命令；② 来自 NM 的当前飞行参数控制模块，其用于分析和发送命令，用机械伺服系统修正飞行参数和一个可移动所有控制面并调节发动机速度的电子发动机速度控制器。

目前，最先进且最广为人知的 OSS 是 Ardupilot。Ardupliot 软件可以在开源硬件和闭源硬件上运行。支持的开源硬件板有 Pixhawk、The Cube（也称为 Pixhawk）、Pixracer、Pixhack、F4BY、Erle – Brain、OpenPilot Revolution、Beagle Bone Blue、PXFmini RPi Zero Shield、TauLabs Sparky。

在考虑 OSH 时，值得注意的是，它由开放设计工作所设计和提供的物理技术工件组成。OSH 意味着硬件很容易识别，以便其他人可以将其与创客运动联系起来。硬件设计如机械图纸、图表、物料清单、PCB 布局文件、HDL 源代码、集成电路布局文件和驱动硬件的软件都可以免费发布，并可供社区访问。

对于开源硬件自动驾驶仪，制造商共享所有必要的数据从而可以根据可用的数据构建相同的硬件。如果将它与 OSS 连接起来,就可以使应用广泛且开源的无人机自动驾驶仪获得改良。

GCS 包括固定或可移动设备,用于观察、指挥以及控制无人机。GCS 可以在地面、水上或空中操作,充当连接机器和操控员的桥梁。无人机 GCS 设计具有特定的功能要求。其基本功能如下。

(1)无人机控制——成功控制和驾驶无人机完成任务的能力。

(2)有效载荷控制——从地面操作传感器的能力。

(3)任务规划——帮助无人机操控员规划任务的功能,同时提供有关无人机能力和限制的所需知识的输入。

(4)有效载荷证据分析和广播——从有效载荷向最终用户传播数据的能力。

(5)系统/飞行器诊断——无人机的自动测试程序,保持地面控制系统的有效维护和部署。

(6)操控员培训——培训无人机控制员以及练习任务的设施。

(7)紧急程序,飞行后分析——收集飞行和有效载荷数据以及在飞行完成后对其进行分析的能力。

支持上述自动驾驶硬件和软件部分的最受欢迎的 GCS OSS 是 APM Planner 2.0、Mission Planner、MAVProxy、QGroudControl、Tower 和 AndroPilot。

DAM 模块包括光学 RS 设备,频段从可见波段到近红外(NIR)的机载图像采集系统、热红外(TIR)、微波系统、有源和无源测距仪器。

为了深入了解 MAV 系统的内部设计,图 3.3 展示了一种完全依赖开源硬件设计的先进系统架构。该系统最初可用于实时视觉导航和摄影测量任务[6-9,12]。

在这个特殊的设计中,对于实时无人机应用,需要合理的计算功率。选择 NVIDIA TX2 处理器从而在尽可能短的时间内进行所有必需的计算。Jetson TX2 将 256 核 NVIDIA Pascal GPU 与六核 ARMv8 64 字节 CPU 平台和 8Gbit 内存(具有 128 字节接口的 LPDDR4)相结合。CPU 平台具有双核 NVIDIA Denver 2 和四核 ARM Cortex – A57。Jetson TX2 构建块体积小、重量轻、功耗低,尺寸为 50mm (87mm)、85g,功耗为 7.5W。这些特性使这款信用卡大小的处理器非常适合 MAV 的应用。Jetson TX2 是当今市场上用于自主框架的一类杰出 GPU 支持板卡。其中央计算能力来源 NVidia 处理器,该处理器仅被分配用于图像计算或其他要求苛刻的计算[7-9]。在该处理器中运行的应用程序不应影响飞行控制过程。飞行控制过程由使用不同处理器的飞行控制器(自动驾驶仪)控制。在那个特殊的设计中,来自视觉系统(NVidia 处理器)的命令传递给飞行控制器,当然只有简单的命令如此,这类似于远程用户命令或一些航路点坐标。飞行控制器执行所有飞行计算。组件之间的这种通信通过标准协议进行,具有快速而直接的特点。

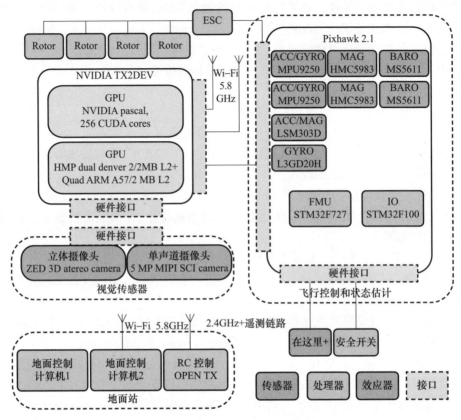

图 3.3　开源硬件 MAV 高级系统设计[6]

3.4　通用消息协议

可以注意到,开源自动驾驶软件可以与许多 GCS 软件和组件一起运行。没有通用的、开源的、可互操作的通信协议是不可能运行的。

MAVlink 的协议是一种如今最流行的且是 MAV 的标准协议,它有一个非常轻量级的消息传递协议,依赖最近的混合式发布-订阅型点对点设计模式来设计机载 MAV 组件,MAVlink 将数据流作为主题发送(发布),而在任务或参数协议上设置的配置子协议则采用点对点重传。这个概念非常类似于机器人操作系统(ROS)[10],后者是另一个著名的开源框架。ROS 是一个灵活的框架,具有一组库、工具和协议,可以使创建跨各种机器人平台的复杂冗余的自动驾驶行为任务变得简单。MAVLink 消息采用 XML 文件,其中每个文件都描述了由特定 MAVLink 系统维护的消息组。大多数 GCS 和自动驾驶仪在 common.xml 中实现此参考消息集(因此大多数语言使用这种方案)。MAV 和 GCS 以及其他 MAVLink 系统使用这些生成的库进行通信[11]。MAVLink 协议非常高效,它的第一个版本(MAVLink 1)每个数据包有 8 字节的开销,以及起始标志和数据包丢弃识别。后续版本(MAVLink 2)有 14 字节的开销,并且更加安全和可扩展。MAVLink 非常适合通信带宽有限的应用,因为它不需要任何额外的成帧。自 2009 年以来,MAVLink 一直允许在许多不同的车辆、GCS 和其他节点以及不同且具有挑战性的通道(具有高延迟/噪声)通信。这是一个非常可靠的协议,因为它安排了探测丢弃、损坏和数据包身份验证需要的方法。该技术可在许多微控制器和操作系统(如 ATMega、dsPic、STM32、ARM7、Windows、MacOS、Linux、Android 和 iOS)上运行,同时支持多种编程语言。其网络上最多可有 255 个并发子系统(如 UAV 和 GCS),它有助于机外和机载通信(如在 GCS 和 MAV 之间,以及在自动驾驶仪和监控摄像机 MAVLink 之间)。高级软件架构的例子(图 3.4)展示了系统的高级结构、软件元素以及它们之间的关系,完全基于开放软件[6]。从任何可用的可能性中选择特定的结构选项,以实现实时计算、兼容的自动驾驶交互可能性,以及任何自定义有效负载集成。

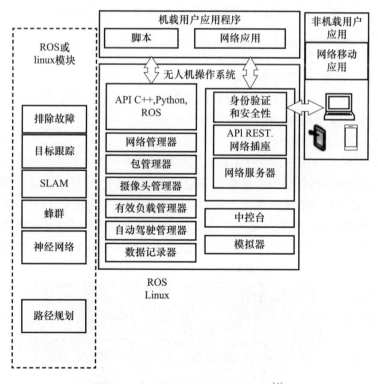

图 3.4　先进的开源 MAV 系统架构[6]

在提出的无人机软件架构上，飞行控制是基于 FC 开源软件（负责飞行控制和单机飞行参数计算）和嵌入式计算机软件。飞行控制软件使用 Ardupilot FC 固件。嵌入式计算单元（NVidia TX2）负责高要求的传感器数据计算。NVidia TX2 正在开发 Ubuntu 操作系统（OS），也是开源的。由于计算需要，一个简单的 Mav-Link 命令会发送到 FC。Ubuntu OS 可以与 ROS 一起用于传感器数据计算。ROS 是一个灵活的框架，用于创建机器人软件，包括各种工具、库，以及简化多方面和稳健无人机行为创建的惯例。在上述软件之上，Flyt OS 集成了 ROS、MAVlink 以及用户应用程序和界面。该操作系统允许将外部 ROS/Linux 库与自定义数据帮助集成。除机外应用外，自定义应用程序可以在任何操作系统和硬件（智能手机、PC、MAC、可穿戴设备、便携式设备）上运行，在无人机和操控员之间建立接口即可。

3.5 GCS 软件

GCS 软件应该能够与所需的自动驾驶仪一起运行，并能够运行和控制所有必要的功能。该软件运行特定的操作系统，包括 Windows、Linux、Android 或 iOS。主要功能如下。

（1）将内置软件（固件）加载到控制特定无人机的自动驾驶仪中。

（2）通过设置、配置和调整无人机操作实现最佳性能。

（3）使用自动驾驶仪以及在谷歌或其他支持的地图软件平台上建立简单直观的点击式 GUI 来规划、保存和加载自主业务。

（4）下载和分析任务日志。

（5）与 PC 飞行模拟器交互来构建完整的硬件在环无人机模拟器[5]。

通过借助附加的硬件，软件应能够通过遥测日志的备份副本监控无人机的运行状态，其中包含有关机载无人机的日志和以第一人称视角（FPV）模式操作无人机的更多信息。且软件必须支持所有自动驾驶功能以及完整的任务规划能力。自主或自动任务的每个方面都可以通过应用程序进行编程和控制（图 3.5）。

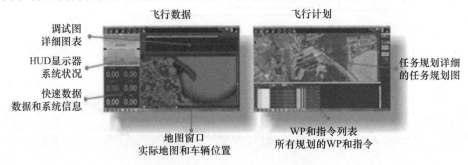

图 3.5　任务规划器的用户界面[5]

目前，任务规划器、APM 规划器、MAVProxy、QGroudControl、Tower 和 AndroPilot 等 OSS 是最常见的软件。这些应用程序是为任何操作系统和几种不同的软件架构而设计的（表3.1）。开源通信协议——MAVlink 是允许使用 Ardupilot 固件进行操作的共同未来。GCS 充当人（操控员）之间的接口，将用户输入转换为 MAVlink 命令，便于无人机的飞行控制器理解。

表3.1 GCS 软件平台

	任务规划器	APM 规划器2.0	MAV 代理	Q 地面控制	塔	Andropilot
平台	Windows, Mac OS X（利用 Mono）	Windows, Mac OS X, Linux	Linux	Windows, Mac OS X, Linux, Android 和 iOS	Android 手机和平板	Android 手机和平板
源代码语言	C#.net	Qt	Python	Qt QML	Android 工作室	Scala（Java）
许可	开源（GPLv3）	开源（GPLv3）	开源（GPLv3）	开源（GPLv3）	开源（GPLv3）	开源（GPLv3）

3.6 处理软件

市场上有很多摄影测量软件，它们是专门为无人机摄影测量和 RS 应用而设计的。最受欢迎的商业软件尽管价格昂贵，但提供了广泛的图像处理工作流程，支持市场上几乎所有的相机和 MAV[7,12-15]。

公司保证提供全面的客户支持、全球服务、培训和分销。软件使用起来相对简单，并且会定期更新。软件有显著的优点，必须在这里提到。MAV 没有等效的开源图像处理软件，可以涵盖所有功能和简单的操作。如今，只有一个开源项目可以交付一些无人机摄影测量产品，如点云、数字地表模型（DSM）、正射校正图像等[12-20]。Open Drone Map（ODM）生态系统就是这样一个案例，其目前有子项目 WebODM 和 Node – ODM。

ODM 是一个 OSS，主要用于处理空中无人机图像。必须强调的是，当前的 ODM 版本号是 0.3.1 bet，并且已经在 2011 年开始。该软件能处理由非度量相机拍摄的微型飞行器图像，生成正射影像、DSM、数字地形模型（DTM）、纹理三维模型和分类点云。ODM 不提供图形用户界面（GUI）来与用户交互，这强加了 Linux 操作系统终端命令的基本知识。准备好的输入图像存储在指定文件夹[9,16,17]中。终端命令开始处理图像并将结果保存在所需的文件夹结构中。ODM 不提供任何额外的查看结果界面，因此需要额外的软件来查看、编辑或评估结果。所有进程都使用中央处理单元(CPU)计算。此软件不支持图形处理单元(GPU)的计算。GPU 计算,尤其是使用 NVIDIA CUDA 技术[18-20]处理图像,与 CPU 计算相比,最高可以加速 8 倍。

因此这个项目被命名为生态系统,额外的子项目支持 ODM。基于 ODM 引擎,用于促进与用户交互的 Web GUI 被添加到这个生态系统中,称为 WebODM。WebODM 旨在成为无人机图像处理的用户友好应用程序。该生态系统还为软件开发人员提供了应用程序编程接口(API)。由于 WebODM 是一种用于 ODM 的基于 Web 的用户界面,因此它生成与 ODM 相同的输出(图 3.6)。

图 3.6　WebODM 用户界面:点云模型(见彩图)

3.7　操控员的信息与通信

在无人机执行任务期间,操控员应该掌握有关平台状态和飞行参数的实际信息。这里有一些处理数据的方式。

视觉上,就视觉视线(VLOS)而言,操控员可以观察飞行平台并控制无人机行为。这是一种主要方法,但必须由操控员进行训练,以便在任何紧急情况下控制无人机。在这种方法中,只有操控员才能拥有平台的相对位置:相对于环境的其他元素和实际平台方向。

在 VLOS 或超 VLOS(BVLOS)操作期间,GCS 飞行参数显示(遥测)帮助。GCS 提供有关无人机状态和实际信息的最全面信息。遥测数据传输到 GCS 和 GCS 屏幕。然而,并非所有数据都对操控员有用,并且甚至有些会干扰操控员的注意力。通常,在双人无人机系统配置中,地面控制系统上散播的无人机数据由

第 3 章　无人机中的开源软件和开源硬件

第二个操控员监控。在这种情况下,当只有一个操控员时,可以使用第三种方法。

飞行数据叠加在操控员的流媒体和屏幕显示(OSD)上。OSD 是叠加在屏幕图片上的一条信息(图像)。飞行操控员的重要信息叠加在流媒体上。这种方法允许操作系统同时监控无人机状态和飞行参数,并控制由无人机传输的视频流。

MinimOSD 及其修改版本如 MinimOSD - Extra 是最流行的 OSS 操作系统。该软件运行在廉价且开放的硬件板上,依赖于 MAX7456 处理器,该处理器是单通道单色 OSD 发生器,不需要外部视频驱动器、EEPROM、同步分离器和视频开关。MAX7456 以 256 个用户可编程字符(NTSC 和 PAL 标准)进入所有国家和全球市场。最小 OSD 额外通过 MAVlink 协议在操控员根据实际需要配置的四个可配置屏幕上接收来自飞行控制器的信息。

在所呈现的示例中,操控员的屏幕(图 3.7)显示了最关键的信息,如固定的 GPS 卫星数量(此处为 0)、方位盘和实际航向(240)归属高度(HA – 1m)(从左上方)。在中间,有一条人造地平线和精确的俯仰和滚转角。从左下角开始,该图显示电池状态(电压、电流和已用电量)、警告部分(此处无 GPS 定位警告)和实际飞行模式。

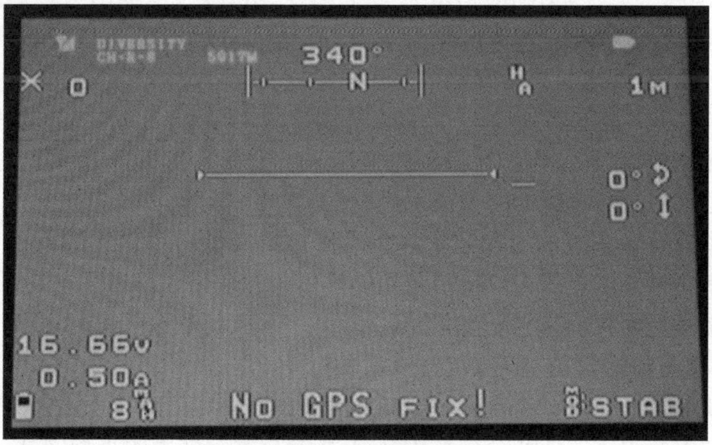

图 3.7　MinimOSD 显示屏

要手动控制或监督无人机,需要无线电链路:地面无线电发射器和相关无线电接收器。无线电控制(RC)发射器正在将操控员的输入更改为所需的无线电信号,并最终改变无人机的机械或电气响应。如今,最先进的开源发射器软件是 OpenTX。准确地说,OpenTX 是一种用于 RC 无线电发射器的开源固件。这个固件由于其高可配置性和比传统专有无线电中的功能更多的功能而广受欢迎。根据 OSS 的开源思想,来自世界各地用户的频繁反馈保证了固件的持续稳定性和质量。

自从 OpenTX 针对主流市场,OpenTX 的固件及其硬件基础在 RC 行业中享有盛誉。OpenTX 项目改进了产品的硬件和软件方面，它们成本低，并且没有大多数制造商对其产品施加营销驱动限制。OpenTX 提供的功能与行业中最高端无线电的功能相匹配甚至超过[19]。与商业方式相比时，最重要的不同之处在于开源社区驱动的固件，与大厂不同，如果需要实现特定功能或用户有很好的改进建议时，几天后就会实现并发布。

3.8 开源平台

由于 OSS 是由其社区编写和共享的，为了使用和传播，它需要一些平台来共享、协作、管理更改和版本更新。GitHub 平台上提供了许多开源项目，这些项目有用于版本控制和协作的代码，并允许人们从任何地方一起处理项目[20]。该平台由用户存储库组成，并提供特定的功能。存储库是必不可少的 GitHub 元素，它包含一个项目文件夹，其中包含所有相关文件以及每个文件的文档和修订历史记录。无论它们是公共的还是私有的，存储库都允许有多个协作者。GitHub 平台主要托管代码。此外，它还支持以下后续格式和基本功能：

（1）问题跟踪（包括功能请求），带有标签、里程碑，以提出与存储库相关的改进建议、任务或问题。对于公共存储库，任何人都可以创建它们，并且由存储库成员对其进行审核。每个问题都包含其论坛、标签和用户分配。

（2）Wiki——存储库上的一个子站点，用户可以在该子站点上直接从 Web 浏览器协作修改内容和结构。在典型的 WiKi 中，文本是使用简化的标记语言编写的，并且经常在文本编辑器的帮助下进行编辑[21]。

（3）带有代码审查的拉取请求和包含用户提交的对存储库的提议更改的评论，以及存储库合作者的相应接受或拒绝。每个拉取请求都有一个论坛。

（4）历史记录可查阅。

（5）文档，包括自动呈现的自述文件。

（6）贡献者、承诺、代码频率、打卡、网络和成员等存储库数据的可视化。

（7）GitHub Pages 是可以在 GitHub 上托管公共存储库的小型网站。

（8）其他数据的可视化，如三维渲染文件、地理空间数据、Photoshop 的原生 PSD 和 PDF 文档。前面提到的所有软件在 GitHub 上都有它的存储库和完整的历史记录。

3.9 未来工作

无人机的大部分挑战都源于在机上和机外执行的用于降低能耗的二分计算

任务,此外,未来趋势是使用无人机,同时通过可靠的无线通信网络来提供物联网设备和物联网服务[22]。

3.9.1 OSH 挑战

OSH 概念涉及一切物理技术,也就是说,它超越了电子技术。取决于人们如何解释这一概念,它可以比 OSS 更受欢迎,因为人们已经探究它很长一段时间了[22]。

技术专长以及对工具和仪器的需求是其中的问题之一。即使程序员的能力很强,他编程也需要计算机、阅读和打字。

OSH 可以真正提升的一种情况是使用 FPGA,因为使用它们进行设计就像使用软件一样朴实无华。然而,与可用的开发语言(如 VHDL、Verilog 和 LabVIEW FPGA)相比,OSH 存在更难以描述和解决的问题,因为它们需要以不同的方式思考挑战。时序和并行化问题无法像串行编程那样学习。

此外,硬件不像软件那样容易复制。如果一个人想要一个 OSS,只需下载它。硬件复制需要一点钱来获取和传输它。OSS 软件可以免费分发,但 OSH(以更易于访问的形式)涉及 VHDL/Verilog/FPGA 代码来以某种方式生成物理系统。

就集成电路和 CAD 项目而言,鉴于今天的限制,生产是完全不切实际的。

硬件描述可以是版权,但不是物理实现本身。真正的物理实现只能通过专利来保护,这需要时间和金钱。OSS 通常可以从版权中受益,因为软件可以与不同于硬件的许可捆绑在一起,故此,不能对外强制执行开源硬件理念[22]。

3.9.2 开放数据

UAV-CPS 迎合了需要不同类型数据库以便对其所有阶段进行基准测试的各种应用程序。然而,处理开源数据(OSD)给其带来了一些挑战。

Open4Citizens 项目[22]制定了促进开放数据使用和支持社区和实践基础的战略。它解决了从广泛过程的组织到网络定义的多层干预的设计复杂性。需要强调的是,现在的趋势是要解耦问题,这样不同背景的人的工作流程才不会受到干扰。

3.9.3 云数据中心

UAV-CPS 的开发需要建立在由类似亚马孙网络服务(AWS)的工具组合提供的云计算基础设施之上。因此,包含分割、特征计算和标记步骤的分析和分类程序要在具有不同大小虚拟机的计算机集群中执行。

必须为 Hadoop JobTracker(又名主节点)保留一个集群节点,只负责任务调度和管理。Apache Hadoop 是一种允许通过简单的编程模型跨计算机集群分布式处理大数据集的软件库。它可以从单个服务器扩展到无数台机器,每台机器都有本地计算和存储。该库不依赖于高可用性高性能硬件,而是在应用层检测和处理故障。因此,高可用性服务是在计算机集群之上提供的,其中每台计算机和每个集群都可能出现故障。随着大数据的增强它变得不可或缺,实现使用它的开源需求无疑也将继续增加:

(1) Apache 软件基金会(ASF)支持许多这些大数据项目。

(2) Elasticsearch 是一个建立在 Lucene 之上的企业级搜索引擎。它的弹性堆栈可以从结构化和非结构化记录中得到理解。

(3) LinkedIn 的 Cruise Control 能大规模运行 Apache Kafka 集群。

(4) 自从谷歌开源以来,TensorFlow 的数量激增。因此,由于其易于使用的概念,它将增加计算智能工具的可用性并普及机器学习。

3.9.4 UAV – CPS 中的众包数据

无人机可以执行人群监视[23]并使用从社交网络中挖掘的信息来促进图像理解。为了评估用例,与机载无人机上的本地视频数据处理相比,无人机可以卸载视频数据让其由移动边缘计算(MEC)节点处理。基于 MEC 的卸载可以帮助节省无人机的有限能量。

在未来的五年,云计算、大数据、软件定义网络/网络功能虚拟化、云安全、传感器网络、分布式机器学习、物联网软件、三维扫描软件等都将比常态增长更快,并发生变化。

3.9.5 无人机群的控制

如今,云服务用于处理大量松散耦合的无人机微服务。向微服务的范式转变要考虑到许多基于云的 UAV – CPS 的假设以及优化服务质量(QoS)和云利用率时的当前机遇和挑战正在不断变化,尤其是当涉及 UAV 群时[24 –32]。使用微服务构建的开源基准测试必须允许以模块化和可扩展的方式模拟和验证全面的端到端服务。

为了整合可以允许群控制和协调与其他更依赖人类的应用程序(如社交网络、娱乐/媒体服务、电子商务、电子银行和不同类型的物联网应用程序)相结合的套件,仍然需要 OSS 和 OSH 发展。这种必要性源于无人机群可能对网络、安全和操作系统等问题产生的影响。它们挑战集群管理和其他有关应用程序设计/部署和编程框架的权衡。由于用户数量非常多,规模影响了真实框架中的微服务,并对性能等所需参数的预测造成了更大的压力。

3.10 小结

本章研究了开(放)源(码)工具(OST)的概念。从开源硬件的角度来看,这一概念也延伸到了 MAV 的情景。OSS 和 OSH 可以适应广泛的应用和改进技术。本章以 OSS 的序言开始,描述其推理方法,建立 OSS 和 PS 之间的对比。OSS 可以使 UAV 领域的很大一部分受益,其目的是推动个人使用 OSS 而不是 PS,这有助于学术、业余和研究社区。随后,本章介绍了无人机人员内部使用的一些 OSS,以帮助无人机的各个部分创新以及 RS/摄影测量信息后准备链。可以想象制造完全自主和可操作的无人机依赖于 OSH 和 OSS。

本章重点介绍与 OSS 和 OSH 相关的问题的多学科性质。它概述了致力于团体发展的不同观点[33]:虚拟结构、集体知识、刺激、创新、共享智能、社区组织、学习和成就。

OSH 需要物理设备,而 OSS 涉及虚拟实体。因此,OSS 产品几乎可以无限制地复制并直接共享。OSH 需要实际成分和费用来复制每个单元或实例。此外,设计硬件具有额外的必要自由度,与全是代码的软件设计相比,这并不是简单的相互替换。

参考文献

[1] Blokdyk, G. "Red Hat Ansible A Complete Guide," 5STARCooks, Plano, Texas, USA, 2019.

[2] P. Iora, M. Taher, P. Chiesa, and N. Brandon, "A one dimensional solid oxide electrolyzer – fuel cell stack model and its application to the analysis of a high efficiency system for oxygen production," *Chemical Engineering Science*, vol. 80, pp. 293 – 305, 2012.

[3] A. D. Team, "ArduPilot." Available at: www.ardupilot.org, vol. 2, p. 12, 2016.

[4] P. Burdziakowski and J. Szulwic, "A commercial of the shelf components for a unmanned air vehicle photogrammetry," *16th International Multidisciplinary Scientific GeoConference SGEM 2016*, vol. 2, 2016.

[5] P. Burdziakowski, "Low cost hexacopter autonomous platform for testing and developing photogrammetry technologies and intelligent navigation systems," 2017.

[6] P. Burdziakowski, "UAV design and construction for real time photogrammetry and visual navigation," *2018 Baltic Geodetic Congress (BGC Geomatics)*, 2018, pp. 368 – 372.

[7] A. M. Coelho and V. V. Estrela, "Data – driven motion estimation with spatial adaptation," *International Journal of Image Processing (IJIP)*, vol. 6, p. 54, 2012.

[8] D. J. Hemanth and V. V. Estrela, *Deep Learning for Image Processing Applications*, vol. 31: IOS

Press, 2017.

[9] B. S. Mousavi, P. Sargolzaei, N. Razmjooy, V. Hosseinabadi, and F. Soleymani, "Digital image segmentation using rule – base classifier," *American Journal of Scientific Research ISSN*, 2011.

[10] M. Quigley, K. Conley, B. Gerkey, et al., "ROS: an open – source Robot Operating System," in *ICRA Workshop on Open Source Software*, 2009, p. 5.

[11] D. Mendes, N. Ivaki, and H. Madeira, "Effects of GPS spoofing on unmanned aerial vehicles," in 2018 *IEEE 23rd Pacific Rim International Symposium on Dependable Computing (PRDC)*, 2018, pp. 155 – 160.

[12] V. V. Estrela and A. E. Herrmann, "Content – based image retrieval (CBIR) in remote clinical diagnosis and healthcare," in *Encyclopedia of E – Health and Telemedicine*, IGI Global, ed. 2016, pp. 495 – 520.

[13] P. Moallem and N. Razmjooy, "Optimal threshold computing in automatic image thresholding using adaptive particle swarm optimization," *Journal of Applied Research and Technology*, vol. 10, pp. 703 – 712, 2012.

[14] B. S. Mousavi and F. Soleymani, "Semantic image classification by genetic algorithm using optimised fuzzy system based on Zernike moments," *Signal, Image and Video Processing*, vol. 8, pp. 831 – 842, 2014.

[15] N. Razmjooy, B. S. Mousavi, M. Khalilpour, and H. Hosseini, "Automatic selection and fusion of color spaces for image thresholding," *Signal, Image and Video Processing*, vol. 8, pp. 603 – 614, 2014.

[16] V. Estrela, L. A. Rivera, P. C. Beggio, and R. T. Lopes, "Regularized pelrecursive motion estimation using generalized cross – validation and spatial adaptation," in *16th Brazilian Symposium on Computer Graphics and Image Processing (SIBGRAPI* 2003*)*, 2003, pp. 331 – 338.

[17] N. Razmjooy, B. S. Mousavi, B. Sadeghi, and M. Khalilpour, "Image thresholding optimization based on imperialist competitive algorithm," in *3rd Iranian Conference on Electrical and Electronics Engineering (ICEEE*2011*)*, 2011.

[18] W. B. aszczak – Ba. k, A. Janowski, and P. Srokosz, "High performance filtering for big datasets from Airborne Laser Scanning with CUDA technology," *Survey Review* vol. 50, pp. 262 – 269, 2018.

[19] T. K. Jespersen, "Software Defined Radio," 2015.

[20] GitHub. (2019). *GitHub Help*. Available at: https://help.github.com/

[21] F. H. Knight, "Capital and interest," *Encyclopedia Britannica*, vol. 4, pp. 779 – 801, 1946.

[22] N. H. Motlagh, T. Taleb, and O. Arouk, "Low – altitude unmanned aerial vehicles – based internet of things services: comprehensive survey and future perspectives," *IEEE Internet of Things Journal*, vol. 3, pp. 899 – 922, 2016.

[23] M. Ghazal, Y. AlKhalil, A. Mhanna, and F. Dehbozorgi, "Mobile panoramic video maps over MEC networks," in 2016 *IEEE Wireless Communications and Networking Conference*, 2016, pp. 1 – 6.

[24] J. Gray. "Design and implementation of a unified command and control architecture for multiple

cooperative unmanned vehicles utilizing commercial off the shelf components." Technical Report, Air Force Institute of Technology Wright – Patterson AFB United States, 2015.

[25] R. Mur – Artal, and J. D. Tardós. "ORB – SLAM2: An open – source SLAM system for monocular, stereo, and RGB – D cameras." *IEEE Transactions on Robotics*, 33, 2017, 1255 – 1262.

[26] A. Bingler and K. Mohseni. "Dual – radio configuration for flexible commu – nication in flocking micro/miniature aerial vehicles." IEEE Systems Journal, 13, 3, 2019, pp. 2408 – 2419.

[27] L. He, P. Bai, X. Liang, J. Zhang, and W. Wang. "Feedback formation control of UAV swarm with multiple implicit leaders." Aerospace Science and Technology, 72, 2018, pp. 327 – 334.

[28] T. Larrabee, H. Chao, M. B. Rhudy, Y. Gu, and M. R. Napolitano. "Wind field estimation in UAV formation flight," in *Proc.* 2014 *American Control Conference*, Portland, Oregon, USA, 2014, pp. 5408 – 5413.

[29] V. Stepanyan, K. K. Kumar, and C. Ippolito. "Coordinated turn trajectory generation and tracking control for multirotors operating in urban environ – ment." In *Proc. AIAA Scitech* 2019 *Forum*, San Diego, California, USA, 2019. https://doi.org/10.2514/6.2019 – 0957.

[30] R. Ahmadi, N. Hili, L. Jweda, N. Das, S. Ganesan, and J. Dingel. "Run – time monitoring of a rover: MDE research with open source software and low – cost hardware." In *Proc. of the* 12*th Educators Symposium* (*EduSymp/ OSS4MDE@ MoDELS*), Saint – Malo, France, 2016, pp. 37 – 44.

[31] R. Mies, J. – F. Boujut, and R. Stark. "What is the "source" of open source hardware?" Journal of Open Hardware, 1(1), 2017, p. 5. https://doi.org/10.5334/joh.7.

[32] Y. Zhang, Y. Gan, and C. Delimitrou. "Accurate and scalable simulation for interactive microservices." In *Proc.* 2019 *IEEE International Symposium on Performance Analysis of Systems and Software* (*ISPASS*), Madison, WI, USA, 2019. https://doi.org/10.1109/ISPASS.2019.00034.

[33] M. R. Martínez – Torres and M. d. C. Diaz – Fernandez, "Current issues and research trends on open – source software communities," *Technology Analysis & Strategic Management*, vol. 26, pp. 55 – 68, 2014.

第4章 基于多重描述编码(MDC)的无人机 MIMO UWB – OSTBC 系统瑞利信道图像传输

正交时空分组码(OSTBC)和多输入多输出(MIMO)通信系统是高性能的新技术,在无线通信中有着广泛的应用。本章介绍了一种 MIMO – OSTBC 无线环境混合结构的多重描述编码(MDC)的 UWB 系统中无人机图像传输技术。MDC 是一种新的图像传输技术,目前尚无相关文献报道。该技术确保在信道错误而导致的分组丢失场景中,可以重建可接受质量且不需要重传的图像。该系统采用不同数量的带有发射天线和接收天线的 UAV 来实现。假设通信信道为瑞利模型,将 MDC 图像传输性能与单描述编码(SDC)进行了比较。试验结果表明,本章提出的混合算法比 SDC 制具有更好的性能。

4.1 引言

无人机技术有着巨大的发展潜力,尤其是近几年来,无人机技术发展迅速。其中一些无人机的应用包括拍摄、娱乐、事故救援、建筑管理、农业和商业无人机货运。这项技术的应用预计使得无人机的数量将从 2015 年的 8 万台增加到 2025 年的 270 万台,大公司希望在农村地区使用无人机[1]。

在紧急情况和灾害(如地震)期间,当通信基础设施受损时,无人机还可用于在运输途中向受灾地区提供无线宽带连接[2]。

这种对无人机的兴趣使人们开始研究具有增强现实(AR)的飞机游戏(AG)、用于链路预算分析的无人机通道以及无人机网络 – 物理系统(CPS)设计[3-5]。最后,UWB 信号可以接收具有出色的临时分辨率的多路径分量(MPC),这可能是开发宽带发行模型 UWB 的一项有趣技术[6-8]。

UWB 的高带宽还可以促进高数据速率、更好地穿透材料以及与用于空地

第4章 基于多重描述编码（MDC）的无人机 MIMO UWB – OSTBC 系统瑞利信道图像传输

（AG）无人机通信的窄带网络共存。

无人机包括无线电控制和自我控制程序[9]。这些无人机通常利用光电能量将信息（如视频和多模态图像）传输到地面控制站，称为高清图像传输系统（HDIT），该系统包含编码、传输、接收和解码，在高清图像处理和计算机视觉（CV）应用中起着至关重要的作用。

特别是，将高清图像和视频数据传输到控制室的应用程序是控制和引导飞行的关键因素，因此需要实时性能。

在光学测量、航空测量、航空摄影等领域，高分辨率图像越来越多地应用于无人机中，以提高观测精度，并且获得了大量的遥感数据。

1080p 高清视频（HDTV）标准以每秒 3 帧、8 字节 YCbCr 的速度每秒产生大约 500Mbit 的数据。2K、4K 视频格式，也用作下一代高清格式，产生了大量的数据。

在大多数应用场景中，无人机需要检查航空图像组件，航空摄影可以实时传输到地面站。为了控制数据传输速率，有必要对图像进行压缩，因此目前无人机领域的带宽都比较低。

此外，无人机需要光伏组件来提供快速移动的特性，它在图像传输系统的功耗、体积和重量方面也非常严格[9-10]。

大多数传统军用飞机使用数据链接技术来创建无线通信系统，研究表明这是一种发送小型无人机图像的新方案。

本章的应用领域从民用空运和航空摄影、资源勘探、电力电缆发现、石油管道监测、消费车辆和无人机视频传输扩展而来。

压缩视频在系统的实现和运行中起到了实实在在的作用，由于技术的发展和系统结构的复杂性，本研究的技术实现能力旨在将图像传输到 UWB 无人机上，且面临着功耗、体积、重量等所有要求。

H.264/AVC 目前仍然很常用，但 HEVC 是流媒体的标准视频编码器。对编码框架和关键技术进行了深入分析。

通常，UAV – CPS 的研究和使用仅限于使用全球卫星导航系统（GNSS）接收机进行制导的外部试验[11]。视觉观测控制最近受到了广泛的空中机器人研究的启发，这些研究是受到了空中机动的复杂性和精度的启发[12]。

低成本和高速的发展使得无人机在紧急情况下的内部应用变得非常有用[13]。超宽带（UWB）通信的进步提供了一个高精度定位，这种高精度定位带来了一个新的应用范围[14]。

超宽带扩频脉冲无线电作为短波通信的替代方案，其概念已经在文献[15]中讨论过，解决多路径回放的能力使其成为室内定位的可行候选方案[16]。

微型无人机技术（MAV）近年来有着广泛的应用前景。此外，无人机航空摄影还可用于三维（3D）测绘[17]、航空测量[18]、精准农业[19]以及搜救[20]和其他有趣的应用[21]。

超宽带(UWB)集中系统的工作方法如图4.1所示。在文献[22]中,超宽带无线电用于估计四架直升机的位置和速度。

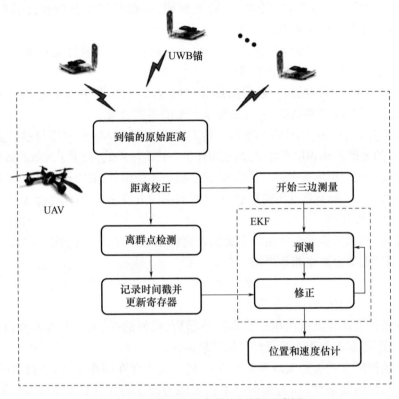

图4.1 基于UWB的定位流程图(见彩图)

四旋翼机上用于测距的超宽带模块用来向锚节点发送测距请求。

文献[23]中提出了用于内部定位的低成本高精度UWB无线电,它允许跟踪仓库中的无人机,以进行可能的自主清点。

所设计的方法是采用锚节点结构的超宽带(UWB)解决方案。因此,它不需要任何有线主干,电池将保持良好的运行状态。

我们开发了一种UWBMAC多技术协议用于无须内置的定位,如Wi-Fi或以太网。图4.2中绘制了建议解决方案的顶层系统图。

图像传输比数据传输拥有更大的数据量和各种价值信息字节[24-25]。无线通信高维图像和视频传输有很多应用,如生物医学无线传感器网络、移动网络、卫星网络、无线传感器网络和应急[25-29]。近年来,多媒体无线通信在通信领域发挥着举足轻重的作用,因此无线图像传输得到了广泛的应用。为了获得更好的图像质量,人们提出了许多技术来提高图像传输系统的性能[30-32]。

第 4 章　基于多重描述编码(MDC)的无人机 MIMO UWB – OSTBC 系统瑞利信道图像传输

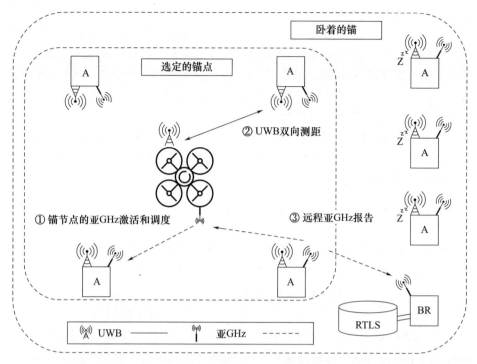

图 4.2　具有电动锚节点的 UWB 定位系统顶层结构图

图像传输系统之一是 UWB,它适用于高维高分辨率多模图像的传输。超宽带(UWB)通信系统是一种高性能、低功耗的新技术,在无线通信中有着广泛的应用。UWB 为短距离高速无线通信提供了一种合适的解决方案。无线电技术可以使用中心频率为 25% 的带宽。功率超宽带系统频带频率低;因为它可以在很宽的频带内工作。该系统对失真和干扰具有较强的稳健性,在接收机中具有较高的对比度和时间分辨率。

低复杂度和低成本是该技术的一些关键优势,加速了其在移动应用中的部署。这些系统中,替代了正弦载波,传输非常窄的脉冲,且系统带宽将达到很多 GHz[33]。跳时超宽带(TH – UWB)技术和直接序列超宽带(DS – UWB)方法是最流行的超宽带方法。

到目前为止,已经为 UWB 系统的图像传输建立了以 TH – UWB 著称的脉冲位置调制(PPM)和采用反模式调制(DS – UWB)的直接序列(DS)方法为基础的跳时(TH)方法[30 – 34]。

关于 AWGN 信道的比较研究有几种类型;在这些系统中,调制 UWB 的性能是不同的。研究人员表明,采用反模式调制的 DS – UWB 系统对 PPM 具有更好的性能和更低的复杂度[35 – 36]。

正交 STBC 代表了良好的直传输模式,它可以处理与经典的最大比合并相同的分集顺序。由于 Alamouti 在[37]中的原始发明以及 Tarokh 和 Calderbank 在[38]中的推广,它们将其用于多输入多输出(MIMO)通信系统。STBC 进行最大似然(ML)解码,可在接收器处执行,但 MIMO 信道可转换为等效标量高斯信道,其响应等于信道矩阵的 Frobenius 范数[39-40]。一些理论建模和模拟[41-42]已经解释了 Nakagami – m 分布[43]和其他情况是瑞利/Ricean 分布。大多数相关研究都使用瑞利衰落或 Ricean 衰落统计模型[44-45]。

MIMO 系统具有更多的宽带带宽应用,并且通过使用 MIMO 天线来实现更大的数据速率。为了在无线通信中获得更多 1Gbit/s 的数据速率,可以使用 UWB 和 MIMO 技术的系统组合[46]。

4.1.1　平坦瑞利衰落信道的效率

QPSK($M=4$)中的非频率选择性瑞利衰落信道的符号错误概率可以表示为下式[47]:

$$P_S^{QPSK} = \frac{3}{4} - \frac{1}{\pi}\sqrt{\frac{E_S/N_0}{2+E_S/N_0}} \cdot \arccos\left(-\sqrt{\frac{E_S/N_0}{2+E_S/N_0}}\right) \quad (4.1)$$

多重描述编码(MDC)技术通过假设数据丢失的风险而适合于无线环境。这些技术用于改进对图像传输的保护,并抵抗通过无线信道传输的数据的损坏[48]。

本章研究采用 QPSK 调制的 OSTBC – MIMO 信道上 MDC 图像的传输。MDC 图像在 MIMO 系统中的传输是一项新的研究,尽管已经开展的研究尚未实施。本章评估了所提出的无线瑞利信道的总体成功率,以将实现的结果与参考文献[49]中的结果进行比较。将通过改变仿真模块及其有效参数,研究了各模块的影响。将通过 MDC 方法和单描述编码(SDC)方法获得的结果进行了比较。

本章结构如下。第4.2节介绍图像 MDC 及其实现。接下来的部分(第4.3节)描述 MIMO 系统。第4.4节和第4.5节分别研究分集和 ST 编码。第4.6节给出了几组不同 t 参数变化的仿真结果,这些变化影响了本章提出的思想的实现。最后,第4.7节给出了结论并讨论了未来工作的展望。

4.2　多重描述编码(MDC)

MDC 是一种基于源代码的技术,它可以保护图像质量不受通道错误的影响。MDC 的主要思想是将原始图像分为四个子图像。图4.3 阐述了此过程,其中每个子采样图像的像素由给定的颜色和形状表示:红色圆圈像素对应于图像版本(IV)1,蓝色方形像素对应于图像版本 IV 2,黄色菱形像素对应于图像版本 IV 3,绿色星形像素对应于图像版本 IV 4。

第 4 章　基于多重描述编码(MDC)的无人机 MIMO UWB – OSTBC 系统瑞利信道图像传输

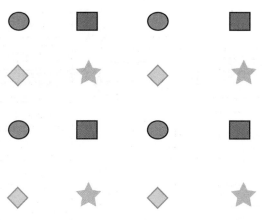

图 4.3　空间域图像的多相子采样(见彩图)

为了测试和验证所提出的系统,对 256×256 摄像图像进行下采样,从而产生 128×128 大小的低分辨率图像,将输入 MCD 模拟系统如图 4.4 所示。

图 4.4　发射器中创建的子图像

每个版本划分为 4×4 个块。然后,用 DC 系数计算每个块的 DCT 变换,然后将这些块的下两个 AC 系数映射为一个向量。因此,该向量输入与量化、信源编码、信道编码、扩频和调制相对应的程序以便通过信道发送信号。在 UAV 接收器部分,操作将反向进行。图 4.5 描述了整个四阶段框架的框图。根据图 4.4,原始输入图像划分为两个或更多版本的数据(子图像)[50]。

每个版本都有令人满意的原始图像质量。如果无人机接收器已经接收到所有的复制,那么解码器将以高质量重建图像数据。否则,信道解码器重建的数据将是低质量的。

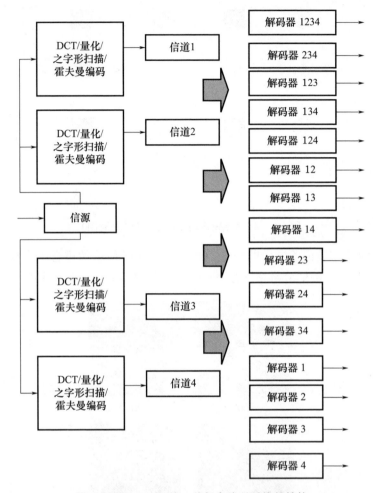

图 4.5　基于 DCT 的四种版本编码系统的结构

解码器显示获得的下采样图像副本。例如,解码器 1234 指示已接收到所有四个版本 1、2、3、4,并且其余解码器呈现版本号。解码输入符号串:$\{X_k\}_{k=1}^N$ 通过四个通道传输,$\{X_{ik}\}_{k=1}^N$ 是对应于第 i 个解码器的字符串。

在该系统中,根据无人机接收器上可用于图像重建的 Ⅳ 版本,总共有 15 种不同的场景。这里,解码器 1234 表示已经接收到全部四个 Ⅳ;解码器 124 指向 Ⅳ 1、2 和 4 的接收;解码器 34 指示已收到 Ⅳ 3 和 Ⅳ 4 的路标;最后,解码器 1 表示只接收到 Ⅳ1[50]。

4.3　多输入多输出

使用 MIMO 天线和 OFDM 等技术有助于最大限度地利用 50Mbit/s 的数据速

第 4 章 基于多重描述编码(MDC)的无人机 MIMO UWB – OSTBC 系统瑞利信道图像传输

率。按照 IEEE802.11n 的建议,要达到 1Gbit/s 的目标,应该使用更先进的技术。UWB 技术与 MIMO 技术的结合可能提供一种解决方案。如果 $NT = NR = 1$,则它称为单输入单输出(SISO)系统,相应的通道是 SISO 通道。频率选择性 SISO 信道的输出可以通过以下方式描述:

$$y[k] = \sum_{k=0}^{L_t-1} h[K,k] \cdot x[K-k] + n[K] \quad (4.2)$$

MIMO 系统是 NI 信号 $x_u[k]$,$1 \leq u \leq NI$ 在每个时刻 k 形成输入并且输出 NO 信号 $y_v[k]$,$1 \leq v \leq NO$,该输入和输出对 (μ,v) 由信道冲激响应 $hv[\cdot]$,$\mu[k,\kappa]$[47] 邻接。图 4.6 说明了 MIMO 系统的结构。

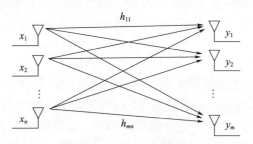

图 4.6 MIMO 系统

$$y_v[k] = \sum_{\mu=0}^{N_1} \sum_{k=0}^{L_t-1} h_{v,k}[k,\kappa] \cdot x_\mu[K-k] + n_v[K] \quad (4.3)$$

随后表示该系统的离散时间模型为

$$\begin{bmatrix} y_1 \\ \vdots \\ y_m \end{bmatrix} = \begin{bmatrix} h_{11} & \cdots & h_{1n} \\ \vdots & & \vdots \\ h_{m1} & \cdots & h_{mn} \end{bmatrix} \begin{bmatrix} x_1 \\ \vdots \\ x_n \end{bmatrix} + \begin{bmatrix} N_1 \\ \vdots \\ N_m \end{bmatrix} \quad (4.4)$$

$$y = Hx + N \quad (4.5)$$

式中:x 是具有发射符号的 n 维向量;N 是 m 维加性高斯白噪声(AWGN)向量;H 是包含零均值复圆高斯随机变量的矩阵;h_{ij} 是从发射天线 j 到接收天线 i 的信道增益[51]。图 4.7 显示了 MIMO 信道的一般结构。

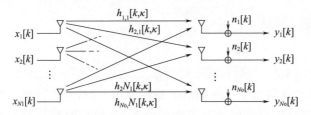

图 4.7 频率选择性 MIMO 信道的一般结构

4.4 分集

MIMO 系统提供分集。由于信噪比(SNR)的平均值存在衰落系数误差，AWGN 信道得到了很大的改善。空间分集的使用包括处理航道破坏性影响的方法。具有时间或频率等多样性类型的组合空间分集可以显著提高系统性能。时空码同时决定了时间和空间的分集程度。

空间分集系统使用多个接收和发射天线阵列。分集增益表示系统可靠性，其中下式的分集增益状态再现了针对 SNR 的错误率演化[52]。

$$d = -\lim_{\text{SNR} \to \infty} \frac{\log(P_e(\text{SNR}))}{\log(\text{SNR})} \tag{4.6}$$

式中：$P_e(\text{SNR})$ 表示在固定 SNR 值下测量的错误率。在 MIMO 系统中，必须使用 NT 发射天线和 NR 接收天线来获得 NT×NR 的最大分集增益。

在 MIMO 系统中，正交空时分组码(OSTBC)编码器用来编码输入符号序列。OSTBC 混合输入信号(来自所有接收天线)，并且信道估计信号以提取关于 OSTBC 编码符号的软信息。

MSE 和峰值 SNR(PSNR)是评价重建图像(处理后)质量的度量，由以下等式给出：

$$\text{MSE} = \frac{1}{M*N} \sum_{M,N} [I_1(m,n) - I_2(m,n)]^2 \tag{4.7}$$

$$\text{PSNR} = 10\lg \frac{255^2}{\text{MSE}} \tag{4.8}$$

其中，前面的表达式假设有 256 个灰度级；M 和 N 分别对应于图像中的最大行数和最大列数；MSE 表示重建图像(I_2)和原始图像(I_1)之间的累积均方误差。MSE 值越小，PSNR 增大时，图像的误差越小。

4.5 仿真结果

该仿真通过 UAV 发送和接收的图像，以及发送和接收该 UAV 结构中基本通信系统信息的步骤说明了所有提出的系统原理，这些基本通信系统也用于传输图像，按顺序运行信道编码调制步骤的任务，然后采取相反的步骤得到最终的图像。

图 4.8 显示了提出的无人机 OSTBC–MIMO 系统的框图。MDC 块获取 256×256 摄像帧，然后对其进行 128×128 子采样处理。然后使用 DCT 对子采样

第 4 章 基于多重描述编码(MDC)的无人机 MIMO UWB – OSTBC 系统瑞利信道图像传输

版本进行量化,将其量化到下一块中以使用作为源编码的算术码,并且在下一块中使用卷积信道编码。

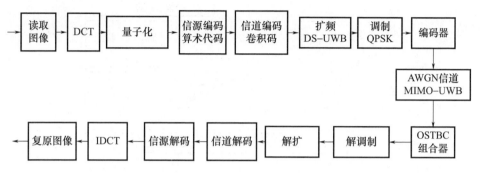

图 4.8 提出的 MIMO 系统框图

使用 QPSK 调制时,假设信道遵循瑞利分布 MIMO,得到信噪比为 15dB 的 2×2 和 2×1 天线的数目。在无人机接收机中,与解码相对应的执行操作发生相反的情况。在下一步的无人机接收机中,对信号进行解调和信道解码,然后进行 IDCT 信源解码。将接收到的具有 128×128 维度的子图像合并后,它们将恢复成 256×256 维度。图 4.9 显示了无人机接收器中图像的最终接收版本。

图 4.9 接收器中获得的子图像

4.6 节将说明天线相互耦合如何影响 MIMO 信道上 OSTBC 传输的性能。图 4.10 显示了 QPSK 调制的 Alamouti 码在有无天线耦合的情况下对每个 SNR 值的模拟。在每次迭代中,通过 MIMO 信道实现 Alamouti 码。

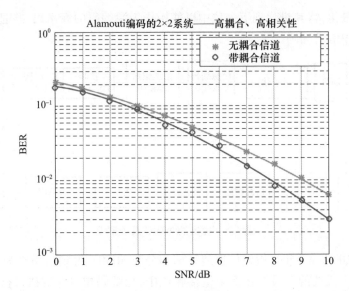

图4.10 在不同的相关和耦合情况下绘制的误码率(BER)–信噪比(SNR)曲线(见彩图)

然后将 128×128 子采样图像合并,以 15 dB 信噪比重建 256×256 原始图像(图4.11)。

图4.11 子图像经 QPSK 合成后的图像

表4.1 给出了使用 MDC 和 DCT 变换的不同丢失版本数的平均 PSNR 值。在版本丢失的情况下,可以通过平均匹配像素来恢复丢失部分的近似副本。

表4.1 PSNR 与不同丢失图像场景的比较

丢失图像数	PSNRSNR = 15dB
0	24.2543
1	23.0673
2	23.6530
3	22.2139

第4章 基于多重描述编码(MDC)的无人机 MIMO UWB – OSTBC 系统瑞利信道图像传输

在表4.2中,仅针对原始版本的 DC 系数说明了一种状态下接收图像的 PSNR。在这种情况下,分析了基于丢失复制数的重建图像的峰值信噪比。

表4.2 峰值信噪比直流系数图像

丢失图像数	PSNRSNR = 15dB
0	20.5103
1	20.4975
2	20.4538
3	20.3125

从 PSNR 的角度可以看出,DC 和 AC 三个系数的图像相对于 PSNR DC 和 AC 三个系数的图像具有更好的质量;因为使用了原始图像的三个系数。图4.12 显示了丢失复制数的 PSNR 值。

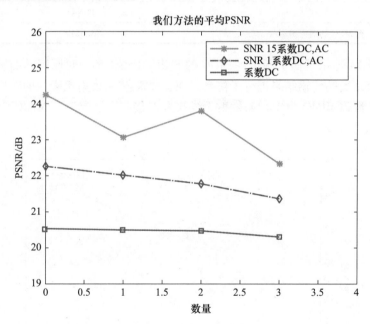

图4.12 不同版本丢失信号的峰值信噪比(PSNR)平均值(见彩图)

图4.13 和表4.3 显示了有无 MDC(SDC)系统的比较。在这种方法中,存在图像丢失的可能性。MDC 方法提供了高质量的图像,避免了 SDC 方法的版本丢失。调制方式为 QPSK,衰落信道,SNR = 15dB,MIMO 为 2×2。不使用 MDC 的 PSNR 方法很少是高值的。

图 4.13 未使用 MDC 获得的图像

表 4.3 无 MDC 的 PSNR 图像

图像	PSNR
无 MDC 的图像摄影师	28.6540

图 4.14 显示了四个版本的 MDC 的 SNR – PSNR 率。对原版本的直流和后两个交流系数的传输结果进行了仿真。在这种状态下，发射机天线的数目是 2；接收天线数为 2（MIMO 为 2×2），发射天线数为 2，接收天线数为 1（MIMO 为 2×1），如图 4.14 所示。

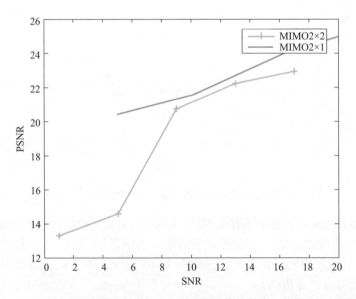

图 4.14 利用离散余弦变换的 MDC 方法的峰值信噪比 – 信噪比（见彩图）

第4章 基于多重描述编码(MDC)的无人机 MIMO UWB – OSTBC 系统瑞利信道图像传输

表4.4 PSNR 图像(MIMO 2 × 1)

图像	PSNR
图像摄影师	23.8765

表4.4 给出了使用 MDC 的图像 OSTBC MIMO 系统中的峰值信噪比,该系统有四个接收副本,尺寸为 128 × 128、衰落信道,SNR = 5,10,15 dB 和 MIMO 为 2 × 1 以及扩频信号。在这种情况下,DC 有一个系数,AC 侧有两个原始副本的系数,这些原始副本是通过由 256 × 256 大小的图像获得的所接收的四个副本的组合来接收的。

图4.15 展示了调制前和解调后获得的12个信号(4个副本的3个系数为12个)的信噪比衰落信道与 BER(误码率)的关系。

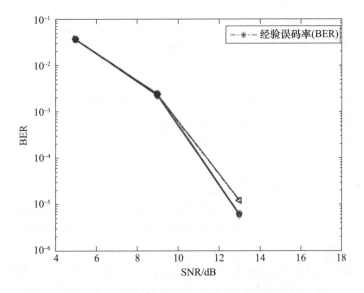

图4.15 AWGN 信道解调后的误码率(见彩图)

在图4.16 中,显示了设置为零的12个信号在信道编码之前和之后的错误率。因此,通道代码用于纠正所有错误。

4.6 讨论和未来发展趋势

UWB 通过单目 SLAM 扩展 UWB 定位,有助于无人机在大规模 UAV – CPS 应用中的顺利使用。因此,无人机室内导航在没有无线定位的区域变得可行。此外,UWB 对性能有可伸缩性影响,它有助于多无人机协同 SLAM,减少计算量。

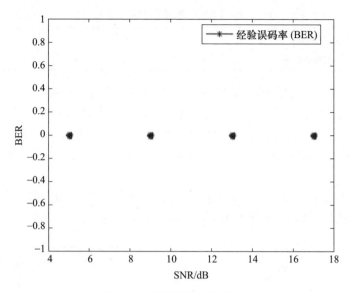

图 4.16 信道译码后的误码率

UWB 可以在稀疏参考节点存在的情况下,利用给定的惯性传感器结构不断优化定位,特别为 UWB 信号的质量评估量身定制[53-54]。

  超宽带技术与带有惯性导航系统(INS)的微型雷达系统一起,可以帮助小型无人机在拒止 GNSS 环境下导航。UWB 和雷达是互补的,在这种情况下,雷达的使用为当前方向提供了足够的证据,而一个独立的 UWB 子系统可能需要大量的基础设施,以便在整个飞行过程中进行充分的位置评估。雷达的速度和超宽带系统的位置可以通过扩展卡尔曼滤波(EKF)等方法与惯导系统集成。惯性测量单元(IMU)可能会由于附加误差而表现出巨大的漂移,因为它可以达到几千米,这就需要 IMU 子系统如 GPS、LIDAR、雷达或 UWB 来抑制误差。此外,UWB 和雷达可以集成,以增强导航性能,从而以较低的基础设施成本实现 GNSS 拒绝环境下的导航[55-59]。

  在包含压缩视频流的通信中,信道错误是一个严重的候选。有几种方法可以减轻这些信道错误,支持 MV 外推(MVE)的运动向量(MV)内插可以隐藏损坏的 H.264/AVC 帧。在基于 UWB 的 UAV-CPS 中,丢失块的 MV 可以通过外推 MV 插值来恢复或进行错误隐藏[60-67]。目前,使用 H.264/AVC 标准[60]在 UAV-CPS 中处理视频是一个挑战。一些令人兴奋的研究,如参考文献[68]试图通过采用带有自适应神经模糊推理系统(ANFIS)控制器的 H.265 编码器来扩展这一点,该控制器包括传输模块、控制部分和具有优良 ER 和 MSE 值的接收级。

  无线可视区域网(WVAN)无处不在,安装无压力,比各种技术更容易访问。WPAN 可以适用低速或高速应用。随着高清视频和其他高端图像及视频通信技

第4章 基于多重描述编码(MDC)的无人机 MIMO UWB – OSTBC 系统瑞利信道图像传输

术的发展，WVAN 可以提供高质量的输出。因此，为了解决这一缺点，将 UWB 技术与 WVAN 技术相结合，以弥补现有的漏洞，从而以优异的速度和直接的接入获得高质量的输出。UWB 具有数据传输速度快、功耗低、价格低廉、传输距离短等特点，是 WVAN 拓扑结构中传输高端图像和视频的首选。超宽带技术可以与环路滤波方法结合使用，以消除输出中的阻塞伪影。如果将计算智能与用于分类目的的环内滤波相结合，则可能以较小的计算负载实现更高的稳健性[61-67,69-71]。

无人机面临的最大挑战之一是安全着陆。从现在起，可视化部署方法将变得越来越流行，因为它们适合在 GNSS 拒绝使用的情况下着陆，特别是当为部署的制导子系统使用地面计算资源并将无人机实时定位反馈给其自动驾驶仪时。在这种具有挑战性的环境下，一个独特的长基线立体框架可以有一个具有广角视场(FOV)的可扩展基线，而不是传统的固定基线结构。此外，除了计算分析外，理论建模还有助于建筑精度评估。超宽带定位网络，连同基于 CV 的导航，需要社区进一步讨论。沿跑道的无源 UWB 锚可以监听来自无人机的 UWB 信号。地面系统可以根据 UWB 锚的几何形状确定目标位置，然后将其发送回无人机。

在无人机 CPS 中使用的所有形式的无线信道，如 RF、光学、声学等，通过雷达和数据通信传输有限的遥感资源。通常，这两个目的是相互冲突的，它们争夺共同的或不兼容的资源。WVAN 应用程序正在快速升级，这要求 RF 融合作为所有涉众发展的先决条件。与这一多方面的僵局相关的充足解决方案空间包含了具有传感和通信能力的 CPS 的协作或代码签名。通过在整个设计阶段牢记无人机 CPS 的一致需求，而不是传播相互干扰的概念，所有系统的组件都可以得到性能增强。UAV – CPS 社区需要为未来企业制定适当的出发设计程序，在概述即将到来的以数据为中心的结构时，考虑到应用程序、拓扑结构、集成级别和最新技术[72-73]。

在体育活动、运动、娱乐等领域的应用要求对 UWB 移动标签的测距精度和准确度进行进一步的研究。这些调查可以证实一些先前获得的结果。有证据表明，天线方向对超宽带测距精度影响不大，而障碍物对测距精度影响较大。精确的 UWB 参数调整和额外的新试验结果会影响测量精度[74]。

4.7 小结

首先，本章介绍超宽带系统。接下来描述了一种基于 OSTBC MIMO 系统的混合结构，用于在瑞利信道中使用 UWB 和 QPSK 调制的 UAV – CPS 图像传输。给出了图像的多描述编码、原始图像的 DCT 分块以及图像的获取方法。所检查的 MIMO 信道是 2×2 和 2×1 信道，并对它们各自的结果进行比较。图像传输依赖于由状态随信噪比差、MDC 和 SDC 的扩散找到的若干选定的规定系数。结果表明,在低信噪比下,峰值信噪比没有变化。研究了几种 MDC 译码方案,结果表明,OSTBC – MIMO 系统在无人机图像传输中具有良好的稳健性。PSNR 结果分别在有 MDC 和无 MDC 的情况下呈现。研究了具有天线选择和 Alamouti 编码的 MIMO 信道。仿真了 Alamouti 系统的误码率 – 信噪比曲线。未来的信道建模工作可以考虑高精度的衰减现象。

参考文献

[1] Tractica,"Commercial drone shipments to surpass 2.6 million units annually by 2025," July 2015,pp.

[2] A. K. A. M. amd A. Tuncer,and I. Guvenc,"Unmanned aerial base stations for public safety communications," *IEEE Vehic. Technol. Mag.* 2016.

[3] H. J. Loschi,V. V. Estrela,O. Saotome,et al. ,"Emergency response cyberphysical framework for landslide avoidance with sustainable electronics," *Technologies* 2018,6,p. 42. doi:10.3390/technologies6020042.

[4] A. C. B. Monteiro,R. P. Franc. a,V. V. Estrela,et al. ,"Health 4.0:applications,management, technologies and review," *Medical Technologies Journal*,2019,2(4),pp. 262 – 276.

[5] T. O. Edoh,"Smart medicine transportation and medication monitoring system in EPharmacy-Net," 2017 *Int. Rural Elderly Health Inf. Conf.(IREHI)*,Lome,Togo,2017,pp. 1 – 9. doi:10.1109/IREEHI.2017.8350381.

[6] F. C. Commission,"First report and order 02 – 48," Apr. 2002.

[7] S. G. I. Guvenc,and Z. Sahinoglu,"Ultra – wideband range estimation:theoretical limits and practical algorithms," in Proc. *IEEE Int. Conf. Ultra – Wideband(ICUWB)* 2008,3,pp. 93 – 96.

[8] S. G. I. Guvenc,Z. Sahinoglu,and U. C. Kozat,"Reliable communications for short – range wireless systems," 2011.

[9] J. Zhao,F. Gao,G. Ding,T. Zhang,W. Jia,and A,Nallanathan,"Integrating communications and control for UAV Systems:Opportunities and Challenges." *IEEE Access*,6,67519 – 67527,2018.

[10] V. V. Estrela,and A. M. Coelho. "State – of – the – Art motion estimation in the context of 3D TV." In:*Multimedia Networking and Coding*. IGI Global, Hershey, PA, USA, 2013. 148 – 173. doi:10.4018/978 – 1 – 4666 – 2660 – 7. ch006.

[11] S. R. N. Goddemeier,and C. Wietfeld,"Experimental validation of RSS driven UAV mobility behaviors in IEEE 802.11s networks," in 2012 IEEE *Globecom Workshops(GC Wkshps)*, Anaheim,CA,USA,Dec. 2012,pp. 1550 – 1555.

[12] M. H. S. Lupashin,M. W. Mueller,A. P. Schoellig,M. Sherback,and R. D'Andrea,"A platform for aerial robotics research and demonstration," *Flying Machine Arena. Mechatronics*,24,

第4章 基于多重描述编码(MDC)的无人机 MIMO UWB – OSTBC 系统瑞利信道图像传输

1,2014.

[13] D. H. S. J. Ingram,and M. Quinlan,"Ultrawideband indoor positioning systems and their use in emergencies," in *Proc. Position Location Navigation Symp. PLANS* 2004,Monterey,CA,USA, Apr. 2004,pp. 706 – 715.

[14] L. W. M. Vossiek,P. Gulden,J. Weighardt,and C. Hoffmann,"Wireless local positioning – concepts,solutions,applications," in *Radio Wireless Conf.* ,2003. *RAWCON'* 03. *Proc.* Boston,MA, USA,Aug. 2003,pp. 219 – 224.

[15] M. Z. W. a. R. A. Scholtz,"Impulse radio:how it works," *Commun. Lett.* ,*IEEE*,Feb. 1998,2, pp. 36 – 38.

[16] R. A. S. J. M. Cramer,and M. Z. Win,"Spatio – temporal diversity in ultrawideband radio," in *Wireless Commun. Network. Conf. WCNC*. 1999 *IEEE* 1999,2,pp. 888 – 892.

[17] F. N. a. F. Remondino,"UAV for 3D mapping applications:a review," *Appl. Geomatics* 2014, 6,pp. 1 – 15.

[18] H. K. Amit Shukla,"Application of robotics in onshore oil and gas industry a review part i," *Robot. Auton. Syst.* 2016,75,pp. 490 – 507.

[19] C. Z. a. J. M. Kovacs,"The application of small unmanned aerial systems for precision agriculture:a review," *Precision Agric.* 2012,13,pp. 693 – 712.

[20] D. S. J. Qi,H. Shang,N. Wang,*et al.* ,"Search and rescue rotary – wing UAV and its application to the Lushan Ms 7. 0 earthquake," *J Field Robot.* 2016,33,pp. 290 – 321.

[21] Y. K. a. I. Nielsen,"A system of uav application in indoor environment," *Prod. Manuf. Res.* 2016,4, pp. 2 – 22.

[22] Z. Q. K. Guo,C. Miao,A. H. Zaini,*et al.* ,"Ultra – wideband – based localization for quadcopter navigation," 2016,4,pp. 23 – 34.

[23] J. B. N. Macoir,B. Jooris,B. V. Herbruggen,*et al.* ,"UWB localization with battery – powered wireless backbone for drone – based inventory management," Sensors,19(3),2019,467.

[24] V. V. Estrela,A. Khelassi,A. C. B. Monteiro,*et al.* ,"Why software – defined radio (SDR) matters in healthcare?" . *Medical Technologies Journal*,3(3),2019,pp. 421 – 9,doi: 10. 26415/2572 – 004X – vol3iss3p421 – 429.

[25] A. Herrmann,V. V. Estrela,and H. J. Loschi,"Some thoughts on transmedia communication," *OJCST*,11(4),2018. doi:10. 13005/ojcst11. 04. 01.

[26] N. Razmjooy,B. S. Mousavi,F. Soleymani,and M. H. Khotbesara,"A computer – aided diagnosis system for malignant melanomas," *Neural Comput. Appl.* 2013,23,pp. 2059 – 2071.

[27] P. Moallem,N. Razmjooy,and M. Ashourian,"Computer vision – based potato defect detection using neural networks and support vector machine," *Int. J. Robot. Autom.* 2013,28,pp. 137 – 145.

[28] P. Moallem,and N. Razmjooy,"Optimal threshold computing in automatic image thresholding using adaptive particle swarm optimization," *J. Appl. Res. Tech.* 2012,10,pp. 703 – 712.

[29] N. Razmjooy,B. S. Mousavi,M. Khalilpour,and H. Hosseini,"Automatic selection and fusion of color spaces for image thresholding," *Signal Image Video Process.* 2014,8,pp. 603 – 614.

[30] H. Z. T. Lv, X. Wang, and X. r. Cui, "A selective approach to improve the performance of uwb image transmission system over indoor fading channels," in *Proc. 6th Int. Conf. Wireless Commun. Networking Mobile Comput. (WiCOM)*, Shenzhen, China, 2010, pp. 1 – 4.

[31] D. A. do Nascimento, Y. Iano, H. J. Loschi, *et al.*, "Sustainable adoption of connected vehicles in the Brazilian landscape: policies, technical specifications and challenges," *Trans. Env. Elect. Eng.* 2019, 3(1), pp. 44 – 62. doi: 10.22149/ teee. v3i1.130.

[32] N. Razmjooy, and M. Ramezani, "Using quantum gates to design a PID controller for nano robots," *Int. Res. J. Appl. Basic Sci.* 2014, 8, pp. 2354 – 2359.

[33] I. Oppermann, and A. M. Hamalainen, "UWB theory and applications," Wiley, New Jersey, USA, 2004, pp. 1 – 7.

[34] M. Z. W. a. R. A. Scholtz, "Ultra – wide bandwidth time – hopping spreadspectrum impulse radio for wireless multiple – access communications," *IEEE Trans. Commun.* Apr. 2000, 48, pp. 679 – 690.

[35] K. K. Z. Bai, "Performance analysis of multiuser DS – PAM and TH – PPM UWB systems in data and image transmission," In Proc. 2005 International Conference on Wireless Communications, Networking and Mobile Computing, Wuhan, China, 2005.

[36] W. Z. Z. Bai, S. Xu, W. Liu, and K. Kwak, "On the performance of multiple access DS – BPAM UWB systems in data and image transmission," *Proc.* IEEE International Symposium on Communications and Information Technology, Beijing, China, 2005, ISCIT 2005, 2, pp. 851 – 854.

[37] S. M. Alamouti, "A simple transmit diversity technique for wireless communications," *IEEE J. Select. Areas Commun.* Oct. 1998, 16, pp. 1451 – 1467.

[38] H. J. V. Tarokh, and A. R. Calderbank, "Space – time block codes from orthogonal designs," *IEEE Trans. Inf. Theory* July 1999, 45, pp. 1456 – 1467.

[39] T. L. X. Li, G. Yue, and C. Yin, "A squaring method to simplify the decoding of orthogonal space – time block codes," *IEEE Trans. Commun.* Oct. 2001, 49, pp. 1700 – 1703.

[40] H. S. a. J. H. Lee, "Performance analysis of space – time block codes over keyhole Nakagami – ," *IEEE Trans. Veh. Technol.* Mar. 2004, 53, pp. 351 – 362.

[41] A. H. H. Shah, and A. Nosratinia, "Performance of concatenated channel codes and orthogonal space – time block codes," *IEEE Trans. Wireless Commun.* June 2006, 5, pp. 1406 – 1414.

[42] A. M. a. S. A. ssa, "Capacity of space – time block codes in MIMO Rayleigh fading channels with adaptive transmission and estimation errors," *IEEE Trans. Wireless Commun.* Sept. 2005, 4, pp. 2568 – 2578.

[43] H. Suzuki, "A statistical model for urban radio propagation," *IEEE Trans. Commun.* July 1997, 25, pp. 673 – 679.

[44] D. I. L. M. Matthaiou, and J. S. Thompson, "A MIMO channel model based on the Nakagami – faded spatial eigenmodes," *IEEE Trans. Ant. Propag.* May 2008, 56, pp. 1494 – 1497.

[45] W. C. Hoffman, Statistical Methods in Radio Wave Propagation, Pergamon Press, Oxford, UK, 1960, pp. 3 – 36.

[46] F. Z. T. Kaiser, and E. Dimitrov, "An overview of ultra – wide – band systems with MIMO," in

第4章 基于多重描述编码(MDC)的无人机 MIMO UWB – OSTBC 系统瑞利信道图像传输

Proc. IEEE Feb. 2009, 97, pp. 285 – 312.

[47] V. Kuhn, "Wireless communications over MIMO channels," *Universitat Rostock*, Germany 2006, pp. 17.

[48] A. R. R. Y. Wang, and S. Lin, "Multiple description coding for video delivery," *Proc. IEEE* Jan. 2005, 93, pp. 57 – 70.

[49] V. S. Somayazulu, "Multiple access performance in UWB systems using time hopping vs. direct sequence spreading," in *Proc. IEEE Wireless Communications and Networking Conf* Mar. 2002, 2, pp. 522 – 525.

[50] M. N. A. Arshaghi, and M. Ashourian, "Image transmission in MIMO – UWB systems using multiple description coding (MDC) over AWGN and fading channels with DS – PAM modulation," *World Essays J.* 2017, 5, pp. 12 – 24.

[51] A. Goldsmith, "Wireless communications," Cambridge University Press, MA, USA, pp. 315 – 320.

[52] H. Jafarkhani, "Space – time coding theory and practice," Cambridge University Press, MA, USA, ISBN: 978 – 0 – 511 – 11562 – 2.

[53] J. Tiemann, A. Ramsey, and C. Wietfeld, "Enhanced UAV indoor navigation through SLAM – augmented UWB localization," in *Proc. IEEE Int. Conf. Commun. Workshops (ICC Workshops)*, 2018, Kansas City, MO, USA, pp. 1 – 6.

[54] J. Tiemann, F. Eckermann, and C. Wietfeld, "ATLAS – an open – source TDOA – based ultra – wideband localization system," in 2016 *Int. Conf. Indoor Positioning Indoor Navigation (IPIN)*, Alcala de Henares, Madrid, Spain, Oct. 2016.

[55] S. Zahran, M, Mostafa – Sami, A. Masiero, A. Moussa, A. Vettore, and N. El – Sheimy, "Micro – radar and UWB aided UAV navigation in GNSS denied environment," In Proceedings of the 2018 ISPRS TC I Mid – term Symposium: Innovative Sensing – From Sensors to Methods and Applications, Hannover, Germany, 2018; pp. 469 – 476.

[56] C. Cadena, L. Carlone, H. Carrillo, et al., "Past, present, and future of simultaneous localization and mapping: toward the robust perception age," *IEEE Trans. Robot.* 2016, 32 (6), pp. 1309 – 1332.

[57] K. Kauffman, J. Raquet, Y. Morton, and D. Garmatyuk, "Real – time UWBOFDM radar – based navigation in unknown terrain," *IEEE Trans. Aerospace Electron. Syst.* 2013, 49 (3), pp. 1453 – 1466.

[58] E. Kim, and D. Choi, "A UWB positioning network enabling unmanned aircraft systems auto land," *Aerospace Sci. Technol.* 2016, 58, pp. 418 – 426.

[59] G. A. Kumar, A. K. Patil, R. Patil, S. S. Park, and Y. H. Chai, "A LiDAR and IMU integrated indoor navigation system for UAVs and its application in real – time pipeline classification," *Sensors, MDPI,* 17(6), 1268, 2017.

[60] H. Marins, and V. V. Estrela, "On the use of motion vectors for 2D and 3D error concealment in H. 264/AVC video," in *Feature Detectors and Motion Detection in Video Processing,* ed. N. Dey, A. Ashour, and P. Kr. Patra, IGI Global, Hershey, PA, USA, 2017, pp. 164 –

186. doi: 10.4018/978 – 1 – 5225 – 1025 – 3. ch008.

[61] J. Zhou, B. Yan, and H. Gharavi, "Efficient motion vector interpolation for error concealment of H. 264/AVC," *IEEE Trans. Broadcasting* 2011, 57, pp. 75 – 80.

[62] M. A. de Jesus, and V. V. Estrela, "Optical flow estimation using total least squares variants," *Orient. J. Comp. Sci. Technol.* 2017, 10(3), pp. 563 – 579. doi: 10.13005/ojcst/10.03.03.

[63] Y. Li, and R. Chen, "Motion vector recovery for video error concealment based on the plane fitting," *Multimedia Tools Appl.* 2017, 76, pp. 14993 – 15006.

[64] T. L. Lin, C. J. Wang, T. L. Ding, et al., "Recovery of lost color and depth frames in multiview videos," *IEEE Trans. Image Process.* 2018, 27, pp. 5449 – 5463.

[65] M. Usman, X. He, K. M. Lam, et al., "Frame interpolation for cloudbased mobile video streaming," *IEEE Trans. Multimedia* 2016, 18, pp. 831 – 839.

[66] D. Chatzopoulos, C. Bermejo, Z. Huang, and P. Hui, "Mobile augmented reality survey: from where we are to where we go," *IEEE Access* 2017, 5, pp. 6917 – 6950.

[67] Y. L. Chen, H. Wang, Y. Hu, and A. Malik, "Intra – frame error concealment scheme using 3D reversible data hiding in mobile cloud environment," *IEEE Access* 2018, 6, pp. 77004 – 77013.

[68] S. M. A. Salehizadeh, "Motion and noise artifact detection and vital signal reconstruction in ECG/PPG based wearable devices." Ph. D. Thesis, University of Connecticut, Storrs, CT, USA, 2015.

[69] D. J. Hemanth, and V. V. Estrela, "Deep learning for image processing applications," *Advances in Parallel Computing Series*, IOS Press, Netherlands, 2017, Vol. 31. ISBN 978 – 1 – 61499 – 821 – 1 (print), ISBN 978 – 1 – 61499 – 822 – 8 (online).

[70] N. Razmjooy, V. V. Estrela, H. J. Loschi, and W. S. Farfan, "A comprehensive survey of new meta – heuristic algorithms," in *Recent Advances in Hybrid Metaheuristics for Data Clustering*, ed. S. De, S. Dey, and S. Bhattacharyya, Wiley, New Jersey, USA, 2019.

[71] N. Razmjooy, and V. V. Estrela, "Applications of image processing and soft computing systems in agriculture." IGI Global, Hershey, PA, USA, 2019, pp. 1 – 337. doi: 10.4018/978 – 1 – 5225 – 8027 – 0.

[72] B. Paul, A. R. Chiriyath, and D. W. Bliss, "Survey of RF communications and sensing convergence research," *IEEE Access* 2017, 5, pp. 252 – 270.

[73] S. C. Surender, R. M. Narayanan, and C. R. Das, "Performance analysis of communications & radar coexistence in a covert UWB OSA system," in *Proc. IEEE Global Telecommun. Conf. (GLOBECOM)*, Miami, FL, USA, 2010. pp. 1 – 5.

[74] T. Risset, C. Goursaud, X. Brun, K. Marquet, and F. Meyer, "UWB ranging for rapid movements," in *Proc. International Conference on Indoor Positioning and Indoor Navigation (IPIN) Nantes*, France, 2018, pp. 1 – 8.

[75] W. Kong, T. Hu, D. Zhang, L. Shen, and J. Zhang, "Localization framework for real – time UAV autonomous landing: an on – ground deployed visual approach," *Sensors*, 17(6), 1437, 2017.

[76] D. Garmatyuk, Y. J. Morton, and X. Mao, "On co – existence of in – band UWB – OFDM and GPS signals: tracking performance analysis," in *Proc. IEEE/ION Position, Location*

Navigat. *Symp*. , Monterey, California, USA, May 2008, pp. 196 – 202.

[77] C. B. Barneto, L. Anttila, M. Fleischer, and M. Valkama, "OFDM radar with LTE waveform: Processing and performance," In 2019 IEEE Radio and Wireless Symposium, RWS 2019 [8714410] (IEEE Radio and Wireless Symposium, RWS). IEEE Computer Society Press. https://doi.org/10.1109/RWS.2019.8714410

第5章 为摄影测量、遥感和计算机视觉记录而生的低空无人机飞行图像数据库

可用航空图像数量的增长刺激了能够从这些图像数据中提取信息的计算工具的研究和发展。然而研发一个计算机视觉软件是十分复杂的,因为有很多因素影响从航空图像中提取信息,如光线、飞行海拔以及获取图像所使用的光学传感器。CV已应用于大多数现代机器中,如自动驾驶交通工具和工业机器人。这项工作的目的是生成一个高质量的低空无人机(UAV)飞行图像数据库,记录其飞行条件,用于摄影测量、遥感和CV。这项工作收集了可见光和热光谱的航空图像,这些图像是在一天的不同时刻、不同海拔以及一年中的不同时间拍摄的。获取图像的摄像头与无人机的自动驾驶仪同步,在试验室中,获取的图像在空间和光谱上都有特征。这项研究向所有社区提供了巴西某一地区的低空航空图像,这些图像中包括精确的飞行信息和捕获图像的信息,以及地面真实情况和地理参考镶嵌图等附加特征。该数据库的使用示例用于无人机自动导航的镶嵌图案生成和CV算法的开发[1-2]。此外,该数据库将作为开发适合图像自主导航的CV算法的基准。

5.1 引言

随着地理数据库中可用图像数量的增加,对高效执行信息提取、操作和存储的工具的需求激增[3-4]。无人机(UAV)和遥感(RS)应用中计算机视觉(CV)工具的开发需要使用适当的图像数据库来发展和评估提出的改进意见[5-6]。

自主图像导航系统用于全球卫星导航系统(GNSS)存在不可接受误差的地区。全球摄影测量和计算机视觉的最新发展使移动地图系统取得了重大进展[7-8]。这些进步使无人机能够在不依赖该地区全球卫星导航系统的情况下执

行任务,以载人飞机飞行员符合规定的方式利用视觉控制飞行,即所谓的视觉飞行规则(VFR)。当使用 VFR 时,飞行员应能从驾驶舱向外看,并根据地面上的参考进行姿态和导航控制。航空学校向有人驾驶飞机的飞行员灌输这种导航方式,因为这种专业知识使飞行员在姿态和导航系统出现故障时能够安全着陆[9]。

通常,无人机都有一个基于 GNSS 的导航方案,因此当这个系统出现故障时,飞机的自主性会降低。对于无人驾驶航空系统(UAS),CV 应弥补因全球导航卫星系统不可用而造成的信息缺失。应该改进视觉里程计、摄影测量和地理参考软件,以实现类似的视觉飞行系统[10]。如参考文献[11]和参考文献[12]的著作中所见,一些研究人员提出了借助卫星图像、同步定位与建图(SLAM)的导航系统。

反复出现的图像导航的挑战包括选择地标、处理大量数据[13]、土地占用变化、气候变化、自然光照变化[14]。需要利用包含目标环境相关特征的相当大的特定数据库来开发自主飞行算法,因为这个问题应该分成几个部分,如选择、检测、地标跟踪、飞机相对位置估计、新的计算和新的修正轨迹。

嵌入式 CV 系统的一个关键开发问题是创建适当的基准数据库,以帮助设计算法和测试,这需要说明使用的光学传感器、采集时的照明以及必须考虑的精确地理位置[15]。

这项工作描述了新的航空图像数据库 AERIALTOUR CAMPUS UNIVAP 的特点,该数据库与其他数据库不同,主要是因为:① 它处理巴西特定地理特征的特定区域;② 它有特定的时间尺度来评估图像中光线的影响。

5.1.1 无人机图像处理系统

图像处理系统的使用在当前的无人机中有所增加,特别是用在目标跟踪系统以及用于校正惯性传感器误差的辅助系统中[16]。

参考文献[17]中的工作讨论了无人机图像处理方法的自主性,包括导航、监视、遥感以及同时处理地图和定位的系统。

对于面向 CV 的无人机应用的开发和测试,需要真实飞行条件下的视频。根据参考文献[23],基于图像的导航系统必须根据运行区域识别不同的特征,如飞越、森林、海洋、沙漠和城市。

前面提到的工作表明,图像处理将是无人机自动化的重要组成部分。此外,军事和民用组织对该领域的研究方向越来越感兴趣,特别是在某些情况或环境下全球导航卫星系统信息被拒绝或无法获得时[23-24]。

最近,数字图像处理技术使用的增长产生了大量用于测试和开发这些技术的数据库。一些作者在图像数据库中编译了集合[25-26]。在费希尔的工作中,他组织了一个图像处理技术和一些图像数据库的集合。普莱斯的研究提供了一个关于计算机视觉(CV)的各种信息的网站。这两个集合都试图将存储库划分为感兴趣的子区域,如图像变换和滤波器、几何图像特征方法、传感器融合、配准和规划方法等。对于遥感,人们拥有由政府维护的大型数据库,如美国地质调查局保存的美国政府数据库[27]。还有一个专门研究的数据库,如美国阿肯色大学所维护的,专门用来解决考古遗址特征的具体问题[28]。

然而,作者没有找到与巴西低空图像相关的具体图像数据库以及飞机的相应飞行日志信息,如纬度、经度、高度和姿态。这些知识对于 CV 算法的发展至关重要,除了实现特定的摄影测量校正外,CV 算法还可以通过图像获得定位和姿态估计算法的精度。

一些关于数据库的工作涉及巴西的特定特征,如[29]研究了米纳斯吉拉斯咖啡区。这个图像数据库来自轨道传感器,只有 64×64 像素的图像,因此很难准确识别地标。在巴西获得的图像是由业余爱好者和旨在营销和传播企业的空中视频获得的,这些图像不包含开发 CV 算法的必要特征[30-32]。这些图像都不能满足地理参考和不同飞行条件下的需要,这是算法开发和测试的必要条件。巴西发现的大多数图像数据库由轨道图像(在 822km 处收集)或高空航空摄影测量(从 600m 开始)组成。在这项工作结束之前,没有无人机在低空(高达 120m)拍摄的图像数据库。

本章提出了一个由无人机拍摄的低空航空图像数据库,名为 AERIALTOUR CAMPUS UNIVAP,其中数据库所使用的摄像头与无人机自动驾驶仪同步。摄像头具有精确的信息,因为它们拍出的照片在试验室中具有空间和光谱特征[33-34]。该系统允许闭环飞行,允许在不同光照条件下的飞行,以及根据不同年中的不同季节采用不同的 RGB 和热图像传感器。这项研究向所有社区提供了巴西一个地区的低空航空图像,以及为捕获而实现飞行的相应信息,以及地面实况和地理参考镶嵌图等附加特征[34-35]。作为使用数据库的一个例子,介绍了四项工作,两项工作涉及镶嵌生成,两项工作涉及无人机自主导航 CV 算法的开发。

本章组织如下。第 5.2 节介绍图像数据库方案。第 5.3 节介绍建立数据库的图像捕获过程。第 5.4 节介绍针对巴西地理的无人机图像数据库的结果和用途。图像数据库的使用在第 5.5 节中进行了研究。第 5.6 节给出了结论和未来的工作。

5.2 航空影像数据库框架

推荐的图像数据库旨在开发和测试用于定位、制图和导航的 CV 算法。因

第5章 为摄影测量、遥感和计算机视觉记录而生的低空无人机飞行图像数据库

此,它必须具有独特的特征,如飞行高度为 30～120m,在封闭回路中的采集路线,不同的土壤覆盖物,在一天的不同时间用不同的光照进行采集,图像的地理参考,以及飞行数据记录。以下部分为其设计和时间标签添加了更多细节。

5.2.1 数据库要求

必须创建一个通用数据库,为了系统地评估针对巴西景观的 CV 算法的准确性和性能。必须牢记以下要求:包含不同颜色、尺寸和形状的空中系统的不同地标;应考虑不同的照明水平;各种气候条件下的航班运行;不同海拔高度的图像采集;用不同的传感器进行图像捕捉;考虑巴西地区的图像采集。

5.2.2 数据库设计

选定的图像采集区域是 UNIVAP Urbanova 校区的一部分。选择这些区域是因为它们有不同的类型,其中包括森林、湖泊、铺砌道路、未铺砌道路、道路交叉口和建筑物。根据飞机的飞行特性,确定了航线的延伸。图 5.1 显示了所选路径。将收集的数据组织成文件夹和子文件夹,用于可见光谱和热光谱的采集,其中每个文件夹对于每次采集都有一个子文件夹。

图 5.1 路径(绿色针脚表示交叉点,按升序编号,以指示路线的方向)(见彩图)

5.3 图像捕获过程

对于图 5.1 所示的第一条航线,选择的拍摄图像的飞机是一架四旋翼机类

型的多旋翼飞机。这架飞机是用 Ardupilot 自动驾驶仪实现的,这是一个小型飞机的开源平台。最初,飞机配置了用于采集 RGB 图像的摄像头,然后对飞机进行了改装,以容纳热摄像头及其部件。

用两个相机捕捉图像:一个用于 RGB 图像,另一个用于捕捉热图像。选择这些相机的原因将会更详细地描述。

佳能 Power-Shot S110 相机的选择是因为它的光学特性:1210 万像素,快门速度间隔为 1~1/2000s。这款相机允许与 Ardupilot 和佳能黑客开发套件(CHDK)集成,这是一个开源项目。CHDK 的目标是增加佳能相机的功能,以便自动驾驶仪能够同步相机拍摄。这对于将自动驾驶信息添加到捕获的图像以及同步多个摄像头至关重要。使用该固件,可以利用相机的 USB 输入来控制拍摄。这种集成使摄像机数据与自动驾驶仪数据同步,并允许与 CV 应用程序接口,如仅获取特定目标图像的应用程序。

这台相机对应于参考文献[33]中使用的那台,分三个部分对其进行了评估。第一部分确定了摄像头对入射电磁辐射每个波长的灵敏度(光谱特征)。第二部分执行辐射测量表征,以确定由光谱学测量的辐射源(球面积分器)的辐射和由佳能 S110 摄像头测量的同一辐射源的彩色信号之间的比率。第三部分根据约翰逊标准,通过计算与目标感知水平相关联的对比度传递函数(CTF),得到空间特征。光谱响应函数决定了三个佳能相机波段(R、G、B)的光谱特征,即入射辐射每个波长的响应。同样,在辐射特性中,对于三个波段中的每一个,相应的数字指定为入射到佳能相机上的辐射。在空间特征描述中,使用了标准的美国空军 1951,并在 CTF 建立,作为探测、识别和识别目标的阈值频率[33]。

在飞行中用来拍摄热图像的相机是 OptrisRPI450 G7 型。选择这台相机的主要原因是它的市场可用性。国家空间研究所的合作伙伴表示对 AERIALTOUR CAMPUS UNIVAP 的兴趣,同意使用这台相机。

它的分辨率为 382×288 像素,最大帧速率为 80Hz(可降低至 27Hz),尺寸为 46mm×56mm×90mm,质量为 320g[36]。该摄像头必须连接到独立解决方案中使用的记录器。相机制造商本身提供了一个名为 PI 轻量级套件的解决方案,该解决方案专用于将在小型无人机中的运输应用[37-38]。

这些特征允许将摄像头结合到四轴飞行器上并采集热图像。

开始为数据库采集图像的目标区域具有气候特征,在凌晨的时候,工作难以展开。由于该地区通常有大量的雾,数据采集时间表设置为当地时间下午 1:00、下午 4:00 和下午 6:00。为了评估光照对算法稳健性的影响,需要在不同的时间进行采集。图 5.2 显示了这些变化的一个例子。

第5章 为摄影测量、遥感和计算机视觉记录而生的低空无人机飞行图像数据库

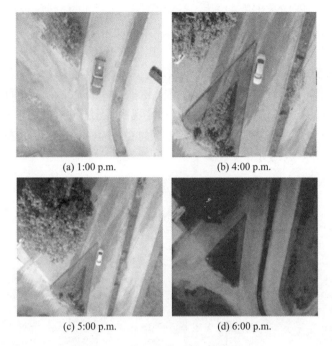

图5.2 光照变化示例(见彩图)

该图像数据库还具有手动注释的地面真实遮罩,可用于自动分类算法的开发。图5.3显示了这个数据库的一个示例遮罩。

图5.3 黏土砖面具(见彩图)

这个遮罩是一个可以放置在地理参考镶嵌图上的图像,用于评估分类软件的性能。在图5.3中,遮罩由代表黏土砖的红色表示。这些遮罩是为面积大于 $9m^2$ 的对象生成的。

为了提高该图像数据库的精度,对地面进行了不同的标记(控制点),以保证基准精度,精度在厘米左右。

为了构建关注区域的镶嵌图,选择了几个支持点。这一程序所涉及的内容远远超过了准确组装瓷砖所需的最低要求。之所以这样做,是因为这个地区将成为自主飞行算法开发的研究区域。有了这些收集的控制点,我们可以在保持宏观区域精度的同时,将这个区域和几个微观区域分开。这些点是用差分校正全球定位系统测量的。有关这些控制点的数据也可在图像数据库中找到。图5.4显示了支撑点的分布。

图5.4 用差分GPS测量土壤中的支撑点(见彩图)

5.4 结果

这项工作的主要成果是一组图像,这些图像具有开发和测试用于摄影测量、遥感和自主图像导航的计算视觉算法所必需的特征。本示例中处理的主要测试用例是自然光条件的变化、不同的地标、飞行高度的变化。第5.4.1节给出了这组图像的一些示例和一些测试用例。

5.4.1 收集的图像

图5.5通过展示一些图像样本说明了使用此图像数据库项目获得的结果。这些图像是无人机在同一航线不同时刻、不同高度和不同气候条件下收集的。AERIALTOUR CAMPUS UNIVAP可用于摄影测量、自动分类、模式识别、环境监测和自主导航等领域的软件开发。

第 5 章　为摄影测量、遥感和计算机视觉记录而生的低空无人机飞行图像数据库

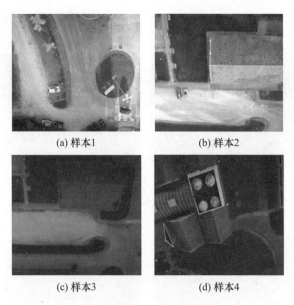

(a) 样本1　　　　　　(b) 样本2

(c) 样本3　　　　　　(d) 样本4

图 5.5　采集图像示例

图 5.5 中的示例显示了自然光的变化,这是评估 CV 算法稳健性的要求之一。自然光提供的另一个关键效果是阴影的变化。图 5.6 显示了阴影变化的情况。

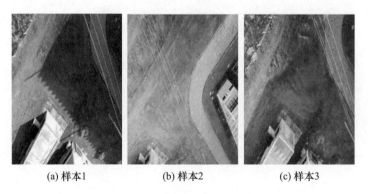

(a) 样本1　　　　(b) 样本2　　　　(c) 样本3

图 5.6　阴影变化示例

这个数据库的另一个关键特性是高度变化,因为每个高度可以从地标中提取的信息是不同的;无人机的高度越高,地标中的细节就越少。图 5.7 显示了此功能的一个示例。

为了描述和比较光学传感器,必须使用标准目标(USAF1951)。这个数据库显示了在标准化目标上不同高度的一些飞行。图 5.8 显示了这种飞行的一个例子。

(a) 样本1 (高度30m)　　　　　　(b) 样本2 (高度40m)

图 5.7　不同高度飞行

(a) 飞越目标　　　　　　(b) 地面目标

图 5.8　飞越标准目标

表 5.1 给出了一些用于图像捕获的航班示例，它打算在不同的时间和情况下继续执行航班，以增加图像库处理的案例数量。

表 5.1　用于图像捕获的航班示例

日期	时间 (p.m.)	高度 /m	LatLong	图数 N
2015 年 6 月 17 日	15:59	20	YES	173
2015 年 6 月 25 日	14:38	20	YES	166
2015 年 7 月 30 日	16:49	20	YES	162
2015 年 7 月 31 日	16:52	30	YES	160
2015 年 7 月 31 日	17:38	40	YES	166

为了更容易地展示所收集的图像，与合作机构一起建立了地理参考镶嵌图，见第 5.5.1 节。

5.5 图像数据库的使用

第 5.5.1 节显示了相应的可见光谱和热图像以及镶嵌图。第 5.5.2 节介绍了使用开发的图像数据库的两项工作:一项工作包括自动检测地标的算法,另一项工作是开发基于地标识别的视觉导航系统,以估计无人机的位置(纬度和经度)。

5.5.1 图像镶嵌

镶嵌图是使用 AERIALTOUR CAMPUS UNIVAP 组装的,以帮助可视化这项工作获得的数据。为了评估组合镶嵌图的准确性,仅使用来自无人机的全球定位系统数据信息来组合另一个镶嵌图,然后与安装有用差分校正全球定位系统收集的地面支持点的镶嵌图进行比较。图 5.9 说明了这种比较,其中黑色标记代表土壤中的测量点,全球定位系统使用差分校正;绿色表示镶嵌图中与地面支持点相同的坐标,红色标记表示仅使用无人机全球定位系统数据的结果。图像镶嵌图发展的更多细节可参见[34-35]。

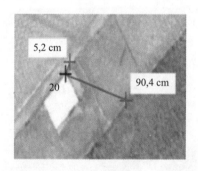

图 5.9 用差分校正 GPS 采集的地面坐标(黑色标记)、用土壤中的支撑点安装的镶嵌图(绿色标记)和仅用无人机 GPS 数据安装的镶嵌图(红色标记)之间的误差(见彩图)

如图 5.9 所示,控制点的添加将生成的镶嵌图的精度提高了 17 倍,将红色标记指示的误差除以绿色标记指示的误差(带有支持点的镶嵌图)。这些在地面上测量的支撑点也保证了第 5.5.2 节中给出的位置估计算法的理想点。

镶嵌装配过程从来自支持点的数据和使用 Pix4D 软件获得的地理参考图像开始。图 5.10 展示了镶嵌结果。

由于飞行器的飞行限制,热镶嵌的面积减小了,用于安装镶嵌图的软件包是 Pix4D 和 Qgis。结果如图 5.11 所示。

图5.10 可见光谱镶嵌图。图片来自参考文献[35]

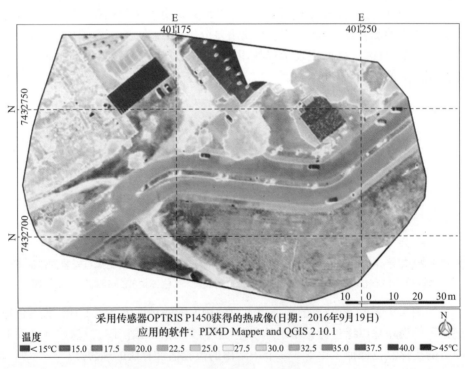

图5.11 热光谱镶嵌（见彩图）

利用图5.10和图5.11所示的镶嵌图，可以通过多光谱成像对该区域进行分析，通过这种方法，可以更有效地识别吸收热量的屋顶，因为它们的轮廓在热图像中有更清晰的范围。

5.5.2 CV 算法的发展

一些工作已经使用所提出的航空图像数据库来测试和验证他们的试验。[39-40]使用 AERIALTOUR CAMPUS UNIVAP 分别开发了自动地标选择算法和视觉导航系统。参考文献[39]建议在飞行路线上自动选择最佳地标。图 5.12 显示了路线部分的结果,其中的标记表示由开发软件自动选择的目标。

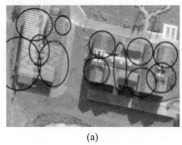
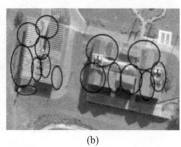

(a)　　　　　　　　　　　　　　(b)

图 5.12　利用参考文献[40]中的球体图像进行航空镶嵌地标选择

第二个已经开发并使用图像数据库的成果是参考文献[41]的工作。它的主要目标是开发一个基于地标识别的视觉导航系统来估计无人机的位置(经纬度)。为此,在工作的两个不同方面使用了相同的图像数据集。第一部分是使用一天中不同时间的航班来测试算法。图 5.13 显示了用于测试的图像示例。数据由两组照片组成,以每秒三张照片的相同频率拍摄,图像分辨率为(4000×3000)像素,每个图像序列是在同一航线上不同时刻的航班上得到的。第一个图像序列是在 2015 年 7 月 31 日 16:30 拍摄的,平均飞行高度为 30m。2015 年 8 月 13 日 17 时 40 分进行了获取第二组图像的飞行,由于是冬季的日落时分,导致了图像变暗,平均飞行高度为 40m[40]。

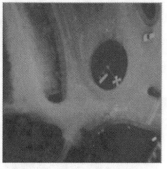

(a) 2015年7月31日拍摄的照片　　(b) 2015年8月13日拍摄的照片

图 5.13　来自 AERIALTOUR CAMPUS UNIVAP 的图像示例

图案如图 5.14 所示。除了显示必要的改进之外，数据集足以显示即使在不同的光照条件下开发的地标识别算法也可生成期望的结果。结论之一是需要实现一种不会降低图像质量的新亮度增强算法。图 5.15 解释了为什么最终的亮度增强算法不够。

(a) 环岛地标　　　(b) 校园工程屋顶　　　(c) 屋顶

图 5.14　从飞行模式图像中选择地标(2015 年 7 月 31 日)

图 5.14 显示了使用包含阴影情况（图 5.14(a)）、结构中的地标（图 5.14(b)）、以及当自然阳光仅到达屋顶一侧时（图 5.14(c)）问题的算法获得的最终亮度模式。

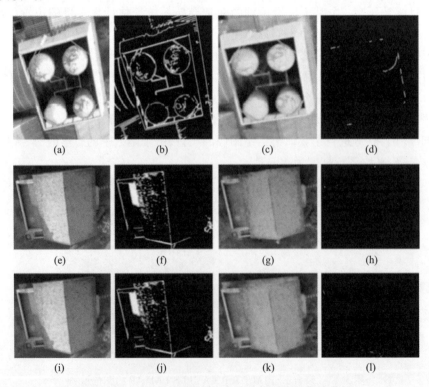

图 5.15　一天中不同时刻飞行中地标识别的候选分析

第二个方面是算法的自动定位测试。由于数据集具有与图像同步的无人机飞行数据,因此,可以估计无人机位置,并将其与全球定位系统/惯性导航系统的导航位置信息进行比较。使用的数据测试图像和布局图案来自于 2015 年 7 月 31 日的飞行,选择了 3 个相同的地标,图 5.16 显示了所得到的结果。

 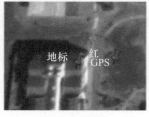 

(a) 环形交叉路口地标　　　　(b) 屋顶的识别　　　　(c) 对屋顶的识别

图 5.16　GPS 自动定位位置比较(见彩图)

(地标识别估计,与无人机在帧中的实际位置相关。红点是地标识别系统估计的位置;
绿点是 GPS 估计值;黄点就是真实的位置)

5.6　结论与未来的工作

在这项工作中,收集的图像有助于 CV 算法的发展,这个数据库处理真实的使用条件,考虑到了不同的自然光照条件和不同的地标。图像数据库必须扩展到尽可能多的飞行条件。

已实施的图像数据库已与其他机构使用,所获得的图像以及一些初步的算法结果已经在图像自主导航、地标识别和自动选择、摄影测量、热分析和多光谱航空分析等领域进行了试验。第 5.5.2[1] 节讨论了其中一些工作。

AERIALTOUR CAMPUS UNIVAP 允许进一步的研究发展。该图像数据库可以扩展增加更多的真实案例,可以增加飞行区域,帮助执行夜间飞行、使用其他传感器捕获图像以及合并不同的飞行高度拍摄到的图像[2,21-22,42-45]。

目前,该数据库已向公众开放,一些合作机构的服务器托管了该数据库。如果需要到目前为止收集的数据,请联系任意一位作者。

在获取高分辨率图像时,超分辨率重建有助于克服传感、硬件、软件和社会等方面的限制[46-48]。特别是,如果随着放大因子的增加,限制条件提供的数据远远不够合适,则可以将新的图像与正在接受审查的站点上的现有知识相结合,丰富一系列结果。众所周知,使用平滑度先验信息可以在一定程度上减轻任务,但是,对于足够大的放大因子,任何平滑度都会夸大结果的平滑度。因此,能够学习描述特定类别环境和场景识别的先验的算法可能会给出更好的随机共振结果。

目标跟踪一直是一个重要而活跃的研究领域。近年来,人们提出了大量的跟踪算法,并取得了一定的成功。然而,用于评估的视频序列通常是不够的,或者是用于评估的视频不太适用于某些类型的算法。许多摄影测量和遥感数据集没有官方的地面真实物体位置或范围,这使得所描述的定量结果的评估存在问题。此外,评估跟踪程序的初始设置和参数不相等,因此,定量效果可能无法比较,有时甚至相互矛盾。为了讨论这些问题,需要对当代在线目标跟踪程序进行广泛的评估,并采用多种评估标准来理解这些方法如何在同一框架内进行。首先,最重要的是将视频序列方面结合起来进行性能检查;第二,可以将大多数公开可用的跟踪器组合到一个具有统一输入和输出格式的代码库中,以简化大规模性能评估;第三,必须对不同初始化环境下的视频算法进行广泛的性能评估。通过分析定量的影响,我们可以发现有效的方法,持续的跟踪和安排潜在的改善前景的研究方向。

管理者和隐私支持者越来越要求通过产品和服务的技术设计以及组织程序和政策来保护机密性[49-55]。对于隐私的研究使得一个新的职业在技术领域诞生——隐私工程师。尽管人们对这种趋势抱有极大的热情,但是很少有人去研究这种隐私保护的新方向是否影响产品的设计,或者如何影响产品的设计。要理解设计是如何用来保护隐私的,需要分析社会技术系统,而不是去文本化的技术制品。亚马孙开发的概念视频提出并解决了关于无人机的公共政策辩论中的隐私问题,这些视频描述了涉及其未来自动无人机包裹递送服务的虚构场景。这里不需要过多地解释、了解特定地区的社会和地理方面是有助于监测和协助该地区,而不会对该地区的生活方式和工作方式造成太大的干扰。借助这个项目和通信技术,人们会意识到,亚马孙的概念视频揭示了人们对隐私问题的更大关注。研究人员需要更多的证据证明隐私问题会影响产品和服务设计,有关这些服务的插图和模拟将有助于塑造消费者对其如何解决隐私问题的期望。虽然这些视频可能不是当前产品的特征,但此类概念视频中产品的角色转变可能会反映在与无人机交付以及其他相关产业链上。随着以消费者为目标的产品功能表现形式的出现,概念视频和其他公开声明一样,如果被扭曲,则可以作为陈述说明的基础。最后,概念视频可以成为有价值的工具,引起监管者和利益相关者根据具体情况考虑何时以及如何进行产品和 UAVCP 设计以保护隐私。

致谢

这项工作得到了巴西航空技术研究所(ITA)、高等研究所(IEAv)、国家空间研究所(INPE)、帕拉伊巴河谷大学(UNIVAP)和技术学院(FATEC)的支持。

第 5 章 为摄影测量、遥感和计算机视觉记录而生的低空无人机飞行图像数据库

参考文献

［1］ Zeng Y, and Zhang R. Energy – Efficient UAV Communication with Trajectory Optimization. *IEEE Transactions on Wireless Communications*, 2017; 16: 3747 – 3760.

［2］ Aroma RJ, and Raimond K. A Novel Two – Tier Paradigm for Labeling Water Bodies in Supervised Satellite Image Classification. In: *Proc.* 2017 *International Conference on Signal Processing and Communication* (*ICSPC*), Coimbatore, India, 2017, pp. 384 – 388.

［3］ Li H, Tao C, Wu Z, et al. RSI – CB: A Large Scale Remote Sensing Image Classification Benchmark via Crowdsource Data. 2017. Available from: http://arxiv.org/abs/1705.10450.

［4］ Yao B, Yao B, and Yang X. Introduction to a Large – Scale General Purpose Ground Truth: Methodology, Annotation Tool and Benchmarks. In: Yuille A. L., Zhu SC., Cremers D., and Wang Y. (eds) *Proc. Energy Minimization Methods in Computer Vision and Pattern Recognition. EMMCVPR* 2007. Lecture Notes in Computer Science, vol 4679, 2007. Springer, Berlin, Heidelberg. Available from: http://www.springerlink.com/index10.1007/978 – 3 – 540 – 74198 – 5.

［5］ Li S, and Yeung DY. Visual Object Tracking for Unmanned Aerial Vehicles: A Benchmark and New Motion Models. In: *AAAI*; 2017, pp. 4140 – 4146.

［6］ Manjunath BS. Aerial Photographs. *Image* (*Rochester, NY*), 1998; 49 (7): 633 – 648.

［7］ Han J, and Lo C. Adaptive Time – Variant Adjustment for the Positioning Errors of a Mobile Mapping Platform in GNSS – Hostile Areas. *Survey Review*, 2017; 49 (352): 9 – 14. Available from: http://dx.doi.org/10.1080/00396265.2015.1104091.

［8］ Toth C, Ozguner U, and Brzezinska D. Moving Toward Real – Time Mobile Mapping: Autonomous Vehicle Navigation. *Proc. The* 5*th International Symposium on Mobile Mapping Technology*, Padua, Italy, 2007, p. 6.

［9］ Ray EL. 2015. Available from: https://www.faa.gov/documentlibrary/media/order/atc.pdf.

［10］ Conte G, and Doherty P. A Visual Navigation System for UAS Based on GeoReferencedImagery. *ISPRS – International Archives of the Photogrammetry*, *Remote Sensing and Spatial Information Sciences*, 2011; XXXVIII1/ (September): 1 – 6. Available from: http://www.isprs.org/proceedings/XXXVIII/1 – C22/papers/conte.pdf.

［11］ Kim J, and Sukkarieh S. Real – Time Implementation of Airborne Inertial – SLAM. *Robotics and Autonomous Systems*, 2007; 55 (4): 62 – 71.

［12］ Karlsson R, Schon TB, Tornqvist D, et al. Utilizing Model Structure for Efficient Simultaneous Localization and Mapping for a UAV Application. *IEEE Aerospace Conference Proceedings*, 2008, pp. 1 – 10.

［13］ Silva Filho P, Rodrigues M, Saotome O, et al. In: Bebis G, Boyle R, Parvin B, et al., editors. Fuzzy – Based Automatic Landmark Recognition in Aerial Images Using ORB for Aerial Auto – localization. Springer International Publishing, Cham, 2014, pp. 467 – 476. Available from: http://dx.doi.org/10.1007/978 – 3 – 319 – 14249 – 444.

[14] Janschek K, and Dyblenko S. Satellite Autonomous Navigation Based on Image Motion Analysis. In: *Proc. 15th IFAC Symposium on Automatic Control in Aerospace*, Bologna Italy, 2001.

[15] Silva Filho PF, Thomaz LA, Carvalho G, *et al.* An Annotated Video Database for Abandoned – Object Detection in a Cluttered Environment. Sao Paulo, Brazil, 2014 International Telecommunications Symposium (ITS), 2014.

[16] Choi H, Brisbane T, Geeves M, Alsalam B, and Gonzalez F. Open Source Computer – Vision Based Guidance System for UAVs On – Board Decision Making. In *Proc.* 2016 *IEEE Aerospace Conference*, Big Sky, MT, USA, 2016, pp. 1 – 5.

[17] Liu YC, and Dai QH. Vision Aided Unmanned Aerial Vehicle Autonomy: An Overview. *Image and Signal Processing (CISP)*, 2010 3rd International Congress on 2010; 1:417 – 421.

[18] Saripalli S, Montgomery JF, and Sukhatme G. Vision – Based Autonomous Landing of an Unmanned Aerial Vehicle. *Robotics and Automation*, 2002 *Proceedings ICRA'02 IEEE International Conference on*, 2002; 3 (May): 2799 – 2804.

[19] Yang K, and Sukkarieh S. Real – Time Continuous Curvature Path Planning of UAVS in Cluttered Environments. In: *Proc. ISMA* 2008. 5th *International Symposium on Mechatronicsand Its Applications*, Amman, Jordan; 2008, pp. 1 – 6.

[20] Fu C, Carrio A, Olivares – Mendez MA, and Campoy P. Online Learning – Based Robust Visual Tracking for Autonomous Landing of Unmanned Aerial Vehicles. In: *Proc.* 2014 *International Conference on Unmanned Aircraft Systems (ICUAS)*, Orlando, FL, USA, ; 2014, pp. 649 – 655.

[21] Razmjooy N, Ramezani M, and Estrela VV. A Solution for Dubins Path Problem with Uncertainties Using World Cup Optimization and Chebyshev Polynomials. In: Iano Y, Arthur R, Saotome O, Estrela VV, and Loschi H, editors, *Proc. BTSym* 2018. *Smart Innovation, Systems and Technologies*, vol. 140. Springer, Cham, Campinas, SP, Brazil, 2019. doi: 10.1007/978 – 3 – 030 – 16053 – 1_5.

[22] Razmjooy N, Ramezani M, Estrela VV, Loschi HJ, and do Nascimento DA. Stability Analysis of the Interval Systems Based on Linear Matrix Inequalities. In: Iano Y, Arthur R, Saotome O, Estrela VV, and Loschi H, editors, *Proc. BTSym* 2018. *Smart Innovation, Systems and Technologies*, vol 140. Springer, Cham, Campinas, SP, Brazil, 2019. doi: 10.1007/978 – 3 – 030 – 16053 – 1_36.

[23] Conte G, and Doherty P. An Integrated UAV Navigation System Based on Aerial Image Matching. *IEEE Aerospace Conference Proceedings*, Big Sky, MT, USA, 2008, pp. 1 – 10.

[24] Sales D, Shinzato P, Pessin G, et al. Vision – Based Autonomous Navigation System Using ANN and FSM Control. In: *Proceedings – 2010 Latin American Robotics Symposium and Intelligent Robotics Meeting*, LARS 2010. 2010, pp. 85 – 90.

[25] Fisher R. CVonline: The Evolving, Distributed, Non – Proprietary, On – Line Compendium of Computer Vision; 2016. Available from: http://homepages.inf.ed.ac.uk/rbf/CVonline/CVentry.htm.

[26] Price K. VisionBib.Com Computer Vision Information Pages; 2015. Available from: http://www.visionbib.com/.

第 5 章 为摄影测量、遥感和计算机视觉记录而生的低空无人机飞行图像数据库

[27] U S Department of the Interior. U. S. Geological Survey; 2017. Available from: https://www.usgs.gov/.

[28] University of Arkansas. Geospatial Modeling & Visualization, 2017. Available from: http://gmv.cast.uark.edu/home/about – the – gmv/.

[29] Penatti AB, Nogueira K, and Santos JA. Do Deep Features Generalize from Everyday Objects to Remote Sensing and Aerial Scenes Domains? In *Proc.* 2015 *IEEE Conference on Computer Vision and Pattern Recognition Workshops (CVPRW)*, Boston, MA, USA, 2015; 44 – 51.

[30] DroneImages. Fotos Aereas Drone Images; 2013. Available from: http://www.droneimages.com.br/fotos – aereas – drone – images.

[31] DRONEFILMAGEMAEREA. FOTOGALLERY; 2014. Available from: https://www.dronefilmagemaerea.com/foto – gallery/.

[32] AltasIMAGENS. Imagens do Alto; 2015. Available from: http://www.altas – imagens.com.br/imagens/.

[33] Almeida RCFd. Avaliacao de Sistemas Eletro – Opticos Imageadores para Missoes de Inteligencia de Imagens na Faixa do Vis. vel. Instituto Tecnologico de Aeronautica (ITA). Sao Jose dos Campos – SP; 2016.

[34] Oliveira LT. Avaliacao do uso de sensor termal a bordo de VANT atraves de analises radiometricas, espectrais, espaciais e posicionais [Master]; 2017.

[35] Nogueira FdC, Roberto L, Korting TS, et al. Accuracy Analysis of Orthomosaic and DSM Produced from Sensors aboard UAV. Santos, SP: INPE; 2017, p. 8.

[36] Optris GmbH. Pi 450 g7; 2017.

[37] Optris GmbH. PI LightWeight Radiometric Aerial Thermography Optris R PI Light Weight; 2017.

[38] Optris GmbH. Operator's Manual Infrared camera. 2017, pp. 1 – 79.

[39] Melo AdS, Silva Filho P, and Shiguemori EH. Automatic Landmark Selection for UAV Autonomous Navigation. In: Cappabianco FAM, Faria FA, Almeida J, et al., editors, *Electronic Proceedings of the* 29*th Conference on Graphics, Patterns and Images (SIBGRAPI'*16*)*. Sao Jose dos Campos, SP, Brazil; 2016. Available from: http://gibis.unifesp.br/sibgrapi16.

[40] Silva Filho PF, Melo A, and Shiguemori E. Automatic Landmark Selection for UAV Autonomous Navigation. In: de Campos Velho HF, editor. *Proceedings of the* 4*th Conference of Computational Interdisciplinary Science (CCIS* 2016*)*. Sao Jose dos Campos – SP; 2016.

[41] Silva Filho PF. Automatic Landmark Recognition in Aerial Images for the Autonomous Navigation System of Unmanned Aerial Vehicles. Instituto Tecnologico de Aeronautica (ITA). D. Sc. dissertation, Sao Jose dos Campos – SP; 2016.

[42] Aroma RJ, and Raimond K. A Review on Availability of Remote Sensing Data. In: *Proc.* 2015 *IEEE Technological Innovation in ICT for Agriculture and Rural Development (TIAR)*, Chennai, India, 2015, pp. 150 – 155.

[43] d'Andrimont R, Marlier C, and Defourny P. Hyperspatial and Multi – Source Water Body Mapping: A Framework to Handle Heterogeneities from Observations and Targets over Large Areas. *Remote Sensing*, 2017; 9:211.

[44] Hemanth DJ, and Estrela VV. Deep Learning for Image Processing Applications. *Adv. Par. Comp.* IOS Press. ISBN978 – 1 – 61499 – 821 – 1 (print) 978 – 1 – 61499 – 822 – 8 (online) 2017.

[45] Wu Y, Lim J, and Yang M. Object Tracking Benchmark. *IEEE Transactions on Pattern Analysis and Machine Intelligence*, 2015; 37: 1834 – 1848.

[46] Jesus de MA, Estrela VV, Saotome O, and Stutz D. Super – Resolution via Particle Swarm Optimization Variants. In: Hemanth J. and Balas V. (Eds), *Biologically Rationalized Computing Techniques for Image Processing Applications*, Springer, Zurich, Switzerland, pp. 317 – 337, 2018.

[47] Deshpande A, and Patavardhan P. Super Resolution and Recognition of Long Range Captured Multi – Frame Iris Images. *IET Biometrics*, 2017; 6: 360 – 368.

[48] Deshpande A, and Patavardhan P. Gaussian Process Regression based Iris Polar Image Super Resolution. In: *2016 2nd International Conference on Applied and Theoretical Computing and Communication Technology (iCATccT)*, Bangalore, India, 2016, pp. 692 – 696.

[49] Wong RY, and Mulligan DK. These aren't the Autonomous Drones You're Looking for: Investigating Privacy Concerns Through Concept Videos, Journal of Human – Robot Interaction, 2016; 5(3): 26 – 54, DOI 10. 5898/ JHRI. 5. 3. Wong.

[50] Estrela VV, Rivera LA, Beggio PC, and Lopes RT. Regularized PelRecursive Motion Estimation Using Generalized Cross – Validation and Spatial Adaptation. In Proc. *SIBGRAPI*, São Carlos, SP, Brazil, 2003.

[51] Estrela VV, Rivera LA, and Bassani MH. Pel – Recursive Motion Estimation Using the Expectation – Maximization Technique and Spatial Adaptation. In Proc. *12th International Conference in Central Europe on Computer Graphics, Visualization and Computer Vision, WSCG* 2004; Pilzen, Czech Republic, 2004, pp. 47 – 54.

[52] Hickey S. Living at Extremes: A Spatio – Temporal Approach to Understand Environmental Drivers for Mangrove Ecosystems, and Their Carbon Potential, D. Sc. thesis, The University of Western Australia, Australia, 2018.

[53] Cissell JR, Delgado AM, Sweetman BM, and Steinberg MK. Monitoring Mangrove Forest Dynamics in Campeche, Mexico, Using Landsat Satellite Data, *Remote Sensing Applications: Society and Environment*, Volume 9, 2018, pp. 60 – 68.

[54] Mueller M, Smith N, and Ghanem B. A Benchmark and Simulator for UAV Tracking. In: Leibe B, Matas J, Sebe N, and Welling M. (Eds), *Proc. Computer Vision – ECCV 2016*. Lecture Notes in Computer Science, 9905. Amsterdam, Netherlands, Springer, Cham, 2016

[55] Martell A, Lauterbach HA, Schilling K, and Nüchter A. Benchmarking Structure from Motion Algorithms of Urban Environments with Applications to Reconnaissance in Search and Rescue Scenarios. *2018 IEEE International Symposium on Safety, Security, and Rescue Robotics (SSRR)*, Philadelphia, PA, USA, 2018, pp. 1 – 7.

第6章 通信需求、视频传输、通信链路和网络化无人机

网络物理系统(CPS)中的无人飞行器(UAV)取决于飞行专用网络(FANET)和计算机视觉(CV)。飞行节点(FN)在无人机网络物理系统中起着至关重要的作用,因为依赖图像对无人机通信构成了严峻而多样的限制。

目前,无人机技术正从单一无人机转变为具有高级目标的协同无人机群。这种情况要求采用创新的网络模式来处理两个或多个飞行节点(FNs),这些飞行节点(FN)可以通过①(在通信范围内没有媒介)直接交换数据,或者②通过类似UAV的中继节点间接交换数据。由于飞行专用网络(FANET)的前提与移动自组网(MANET)和车载自组织网络(VANET)不同,因此设计无人机自组网是一个非常复杂的问题。飞行专用网络(FANET)在飞行节点(FN)的移动性、连接性、数据路由、云交互、服务质量(QoS)、应用程序类型和体验质量(QoE)等方面具有特殊性。

本章将介绍无人机作为特殊节点运行时面临的挑战,并导出无人机的网络模型;并总结了飞行专用网络(FANET)的新兴前景、影响以及它们如何适应多媒体的应用程序。

6.1 引言

目前,在网络物理系统(CPS)中,利用无人机群来提高飞行节点(FN)的通信距离和信息获取能力是至关重要的[1]。在缺少通信基础设施的情况下(如难以到达的地区或遭受自然灾害或人为灾难的地区),除了建立一个强大的团队和地面基站(GS)网络之外,建立一个可靠的网络也变得更加困难。

飞行Ad-hoc网络(FANET)网络层(NL)的主要任务是充当中介,同时使用无人机群组作为移动Ad-hoc网络(MANET)在其级别上执行常规路由。FANET的路由协议在很大程度上取决于两种通信类型:①无人机到无人机

(U2U);② 无人机到基础设施(U2I)。在 U2U 通信中,无人机通信是为了满足不同应用程序的前提条件,如协作路径规划和目标跟踪。无人机可以直接相互通信,也可以通过多跳路径与其他无人机通信。这些相互连接的无人机可以处理短距离或远距离的数据通信传输,而距离的选择取决于必要的数据传输速率。

无人机对基础设施(U2I)的连接包括无人机和固定基础设施(如卫星和地面基站 GS)之间的数据交换,以便通过网络提供服务。此外,通过无人机进行广播,而 U2I 也会有一些警告。

无人机和地面基站服务系统(GS)之间的网络基础设施通信的安装和维护成为需要解决的主要问题,主要是在具有不同有效载荷容量、传感器、成像设备、航空电子设备、通信范围和飞行续航能力的异构飞行节点(FN)的情况下(图 6.1)。

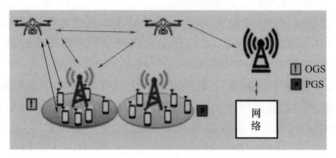

图 6.1　无人机作为网络网关(见彩图)
(OGS:过载的 GS;PGS:有问题的 GS)

6.2　飞行自组织网络

近年来,由于 UAV-CPS 的飞行节点(FN)具有很强的移动性,采用自组网方式的多用途无人机对网络层提出了更高的要求。因此,这一层主要是放置在空中航线(Aerial NL)的亚层中的 VANET 子类。

网络层(NL)指的是主要负责路由的部分,它通过不同的路由器/节点选择最佳的路径连接发送端和接收端节点。FANET 主要作为这些层之间的中介,同时在其层上进行常规路由[2-3]。

FANET 具有以下优势[2]。

- 减少任务结束时间:可以更快地执行侦察、监视、人群保护、搜索和救援等任务。
- 降低总维护成本:与大型昂贵的无人机相比,一些小型无人机的成本更低,而且更容易购买和维护。
- 提高可伸缩性:它毫不费力地通过插入新的飞行节点(FN)来扩大操作区

域。UAV-CPS通过允许新添加的无人机动态地重组FN的路由表。
- 提高生存能力:多用途无人驾驶飞机对硬件/传感器故障更具弹性。当传感器/执行器出现故障,或者FN失控时,剩余的无人机可以继续执行任务。
- 降低可探测性(短雷达截面):微型无人机由于其尺寸和复合结构而具有低雷达截面、小红外特征以及低声学特征。因此,雷达(尤其是与飞机和大型无人机相比时)无法轻易检测到它们。

多用途无人驾驶飞机具有显著的优势,但考虑到它们的动态网络拓扑,两个远距离FN的通信仍然是一个繁重的课题。据收集到的文献表明,一些FANET的路由协议依赖于FN的通信特性。在FANET中,主要有两种不同类型的通信:无人机到无人机(U2U)和无人机到基础设施(U2I)通信[2-4]。

在无人机到无人机(U2U)通信中,各无人机之间相互交换信息,以满足不同应用领域的需求,如协同路径规划、目标跟踪等。两个无人机可以直接相互对话,或者也可以在其他无人机上创建多跳路径。无人机之间可以进行短程或远程通信。范围选择还是取决于预期的数据传输速率[2-4]。

在无人机到基础设施(U2I)通信中,无人机与固定基础设施(如附近的地面基站服务系统、卫星或陆地车辆)进行通信,为全球网络中的其他运营商提供服务。无人机和无人机到基础设施(U2I)之间的通信也是一个很难解决的问题(图6.2)。除了传感器之外,还可以使用其他类型的天线来提高数据传输速率和UAV-CPS性能。可以在FANET[2-4]中使用GPS接收器和定向天线(类似于相控阵天线)有效地建立通信链路。

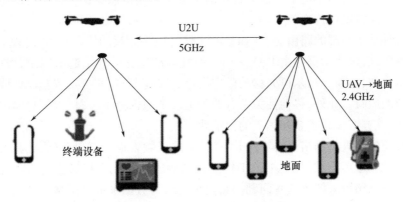

图6.2 无人机和U2I之间的通信[5]

6.3 FANET 协议

由于FN的高移动性和FN之间的距离不断变化,FANET的链路质量也会发生波动。由于链路质量波动和故障,FANET MAC 设计将会面临新的考验。同时,延迟也会使问题复杂化。定向天线有助于增加不同场景下的通信范围、空间复用和增强安全性[6]。

有研究提出了一种自适应 MAC 协议[7],该协议具有:① 用于数据包传输控制的全向天线;② 用于数据包的定向天线。使用这种方法可以改善端到端的延迟、吞吐量和误码率(BER)。出现了一种基于指令的方法来更新目标数据[8],除了处理由于过度移动而导致的链路故障之外,还可以克服普通的基于竞争的协议中的故障。具有多包接收(MPR)功能的全双工无线电可以提高多用途无人驾驶飞机网络中的 MAC 性能。全双工系统减少了延迟,因为每个节点可以同时发送和接收,并且具有多包接收(MPR)功能则会加强多用途无人驾驶飞机的吞吐量。

文献[9]中改进的优化链路状态路由(OLSR)协议依赖于定向天线。在 OLSR 协议中,关键步骤是选择多点中继(MPR)。然而减少 MPR 的数量将导致传输的控制数据包更少。使用目的地信息来传输数据包[9],如果与源的目的地距离小于定向天线最大值的一半,则使用 DOLSR。否则,OLSR 完成路由。这项工作还提出减少 MPR 的数量,从而减少控制开销,从而减少延迟并提高整体吞吐量。

时隙预留安排与 AODV 一起使用,以减少冲突。参考文献[10]中的研究工作提出了一种混合方法来最小化中间节点通信。这种时间预留策略类似于时隙 ALOHA,在时隙 ALOHA 中,其中每个节点都接收一个时隙,以在该特定时隙内向主节点或具有与其他节点通信特权的集群头传输信息。这种方法有助于减少冲突和提高数据包传递效率。

一种基于地理的路由协议(如参考文献[11]中建议的协议)可以发现下一个最佳可用跳,以成功消除由 UAV – CPS 的极端动态移动性引起的间歇性连接的影响。最初,高斯 – 马尔可夫移动模型促进了飞行节点(FN)位置预测以降低路由失败的风险。此外,移动性关系有助于更准确地选择下一跳,以便更精准地进行路由,该方案提高了簇和簇头的稳定性。

目前的无人机网络聚类算法并不适用,除了频繁地集群更新外,还具有较高的移动性。参考文献[12]中的工作通过依赖于使用无人机属性的集群加权模型的移动性预测来解决这些问题。它使用树作为数据来估计网络拓扑结构足以进行字典预测,并支持用于调节链接到期时间的移动模型。由于采用了簇头选择和按需维护策略,它支持更稳定的簇结构和更高的网络性能。FANET 的显著特征加强了独特的设计关注点,如下所述。

1)适应性[4]

(1)在多用途无人驾驶飞机操作期间,FANET 多个参数可能会有所不同。FN 的移动性很强,并且总是会修改它们的位置。无人机的路线可能会有所不

第6章 通信需求、视频传输、通信链路和网络化无人机

同,并且由于操作要求,无人机之间的距离不可能恒定。

(2)在执行任务期间,飞行节点(FN)可能会因为技术事故或对多无人机系统的攻击而失败。虽然无人机故障减少了无人机的数量,但在改变FANET参数的同时,可能需要其他策略来保持多用途无人驾驶飞机的运行。

(3)气象条件也可以干扰FANET,因为如果天气突然变化,FANET的数据链接可能不会持续存在。FANET应设计成在高度动态的大气中继续操纵。

(4)在多用途无人驾驶飞机操作期间可能更新分配。而新的任务信息可能导致飞行计划更新。例如,在多用途无人驾驶飞机搜救行动中,新的情报报告到来后可能发生变化,使任务集中在特定区域,从而影响飞行计划和FANET参数。

(5)FANET设计应根据任何变化或故障进行自我调整。FANET物理层应根据节点密度、节点间距离和环境变化进行自我调整。它可以扫描参数并选择最佳物理层替代方案。FANET环境的高度动态特性也会影响NL协议。Ad-hoc网络路由维护与拓扑变化密切相关。因此,无人驾驶飞机性能取决于路由和链路更改协议的适应性。传输层也应符合FANET的状态。

2)可扩展性[4,13]

(1)与单个无人驾驶飞机相比,协作型无人机的工作可以增强整个无人驾驶飞机性能,这是采用多用途无人驾驶飞机的主要动机。在许多应用中,性能随着无人机总数的增加而增长。例如,更多的无人机可以更快地完成搜救行动。FANET协议和算法必须允许任意数量的飞行节点以最少的可能的性能下降。

(2)监视服务器直接与操控员协作,设置任务并提供有关任务的必要信息。用户通过定义监视区域和要部署的飞行节点(FN)数量来协调任务。当操控员输入调查区域和应部署多少飞行节点时,服务器使用飞行节点放置算法估计单个飞行节点的放置。该服务器为无人机群完成有用的关键的任务。首先,如果由服务器控制无人机的旋转,则可以克服飞行时间短的限制。当飞行节点电池电量不足时,服务器读取其电源状态并认为飞行节点有故障。所有有故障的无人机都将从无人机机群中移除,并由另一个模块进行交换,以防任何其他的无人机进入。否则,布局算法以高优先级优先的方式重新定位FN。布局算法在GS上运行是因为GS由于没有重量和大小限制而具有更高的计算能力。

3)延迟[14-15]

(1)延迟是FANET的一个重要设计关注点,延迟需求完全取决于应用程序场景。特别是对于实时的FANET应用程序,如监控,信息包必须在特定的延迟范围内提供,多用途无人驾驶飞机的避障也需要低延迟。

(2)基于FANET对IEEE 802.11进行了单跳包延迟调查,每个FN都建模为一个M/M/1队列,平均数据包延迟通过分析获得。仿真分析证实了这一原理,并根据仿真分析记录给出了出色的数值结果,其中数据包延迟遵循Gamma(伽

玛)分布。Zhai 等研究了 IEEE 802.11 传统无线局域网的数据包延迟性能,采用指数分布随机变量来近似 MAC 层数据包服务时间。本章还证明了 manet 和 fanet 的包延迟行为是不同的。MANET 协议可能不能满足 FANET 的延迟要求,因此延迟敏感的无人机群需要新的 FANET 协议和算法。

4)无人机平台约束[17]

(1)无人机重量影响 FANET 通信硬件性能。更轻的硬件意味着更轻的有效载荷,延长了无人机的续航时间,并允许在无人机上安装额外的传感器。如果总有效载荷是恒定的,通信硬件很轻,那么就可以携带更多的尖端传感器、驱动器和其他外围设备。

(2)空间限制是 FANET 设计的另一个限制,因为它会影响可安装到无人机平台上的通信硬件。

5)带宽要求[18]

(1)大多数 FANET 应用程序的目的是收集环境数据,并将获取的信息发送到地面基地进行监测、监视或救援行动。与目标区域相对应的可视数据必须在无人机和指挥控制中心之间使用严格的时延限制进行中继,这需要高带宽。此外,利用先进的传感器可以获得高分辨率的视觉信息,这使得对带宽的要求更高。多个 FN 的协调也需要额外的带宽。

(2)相反,存在各种带宽使用限制,如通信信道容量、无人机的速度、无线链路易出错结构、广播安全性不足等。FANET 协议必须满足带宽容量前提条件,以便在各种约束条件下中继高分辨率的实时图像或视频。

6.4　FANET:流媒体和监控

实时传输视频流对带宽、延迟和损失提出了严格的要求,以保证连续的视频播放[19-20]。由于视频的高编码率,如果所需的吞吐量保持在可用的网络容量下,则可以保证连续的视频播放。视频数据包接收中的高丢包或大延迟会导致视频播放失真或频繁暂停。我们现在讨论使用无人机进行各种应用的视频流的现有示例[21]。

参考文献[22]研究了一种使用无人机的实时空中交通监视系统。使用基于中继的网络体系结构来流式传输来自无人机的视频。无人机使用模拟传输将视频流传输至 GS(如笔记本电脑),GS 使用移动宽带蜂窝网络通过互联网中继视频流,假设其在附近可用(图 6.3)。最终用户可以近乎实时地通过互联网上传视频。然而,问题是网络拥塞和无线链路波动,这不能保证只允许低质量视频流的专用带宽量。现场试验表明,如果不使用存储服务器,流畅的视频只能以

45kB/s 的编码速率、15 帧/s 的帧率和 160/120 的屏幕分辨率传输。当使用存储服务器时,它可以传输和存储来自互联网的视频,并且编码速率增加到 109kB/s 是可以承受的[21]。

研究在 IEEE 802.11 自组织网络上使用 AR 无人机的视频流,目的是监控农业区域[23],否则可能需要更多的时间和人力。设计的控制软件可以估计通信范围和视频传输速率。当距离在 40~74m 之间变化时,对视频流进行的测试结果是平均视频速率为 700B/s,而在两跳场景(高达 200m)中,平均吞吐量为 612B/s。我们建议,在多跳场景中使用具有 QoS 支持的跨层解决方案和路由协议,可以改善视频质量[21]。

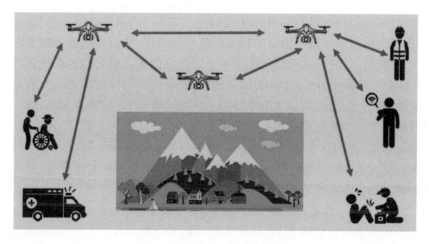

图 6.3 无人机作为中继节点

荒野搜索和救援(WiSAR)是基于视觉的空中搜索。它使用的是翼展为 42~50 英寸(1 英寸 = 25.4mm)的小型无人机[24]。一个 900 MHz 的收发器用于数据通信,而模拟 2.4GHz 发射器用于视频流。为了寻找失踪人员,需要考虑最后出现的地点和旅行方向等标志。视频图像具有 640/480 分辨率。研究发现,如果无人机的工作高度保持在 60~40m 之间,就有可能探测到衣服和人造物体的异常颜色。现场测试表明,由于分心抖动、定向旋转、噪声和失真的图像增加了搜索和检测过程的难度,因此需要提高视频质量。计算机视觉(CV)算法用于增强视频特征的稳定性和时间局部性[21]。

例如,火灾、洪水、飓风、地震等灾害情况需要灾后评估、规划和响应[25]。在这种情况下,无人机可以为灾害管理人员提供安全、及时和关键的信息,以规划救援行动。传输图像数据需要 65kB/s 的下行链路[25],而通过无线局域网传输 H.264 或 MPEG-2 视频则需要大约 2MB/s 的编码速率[26]。

尽管无人机视频流的真实示例不多,但是基于现有示例,可以知道高质量视频流需要 2mB/s 的编码速率。除了视频速率的要求外,控制和遥测通信数据也因应用程序而异。考虑到现有技术,可以说当前的非授权技术在需要大的通信范围和基于宽带授权的技术不可用的情况下,可能无法满足多样化视频流应用的需求[21]。

6.5 讨论和未来趋势

6.5.1 FN 的布局搜索算法

更重要的是目标区域和 FN 的物理耦合。如果一个区域与分配的 FN 完全解耦,那么无人机可能不得不为每一帧飞行很长的距离,从而导致更多的监视停机时间。该算法设计支持目标监视区域和分配给它的无人机之间的物理耦合。由于 FN 的优先级得分和无人机 ID,即使它移动,分配到目标区域的无人机也将保持分配到相同的目标区域。该策略保障了目标区域和 FN 的物理耦合,除非群体成员发生变化或优先级顺序发生变化。

配备摄像头的 FN 的部署在监视计划中起着关键作用。对于一个移动监视基础设施有太多 FN 的情况下,自动部署算法要比人工部署优越得多,而人工部署还需要人力。布置尽可能多的 FN 将不断提高监视的效率,但当无人机数量有限时,有效的 FN 布置是获得最大效能的基础。因此,可以设计新颖的算法来根据用户提供的要求自动部署 FN 以形成监视框架。布局算法[27-30]需要分数或指标,它包括以下一些步骤。

(1)每个区域都有一个重要程度,不同的区域的重要性需要不同程度的关注。区域状态可以用得分来表示,得分涉及几个问题,例如与图像分辨率成正比的相机质量的好坏。

(2)为 FN 寻找最佳的三维摄像机放置的位置需要大量计算,这会随着高分辨率图像的增加而大大增加。每个地方都需要用所有相交摄像机和对焦区域的得分总和(合适的指标)来评估。通过计算智能(CI)可以获得计算工作量较小的布局方法[31-32]。新的摄像机放置需要搜索位置、要覆盖的区域和无人机的尺寸。

(3)迭代无人机的布局算法并不能保证 FN 的全局最优布局。全局最优解需要更大范围的搜索,这是不现实的。更好的次优解决方案可以通过在固定数量的 FN 上使用一些合适的指标进行多次迭代,或者通过 FN 基于性能的排序,并尽可能专注于高优先级领域来产生。如果 FN 的数量不够,那么对于任何目标区域都没有空闲时间。经过一轮优化的放置迭代后,每个 FN 相机都会拉伸以适应未覆盖区域的大小。此外,如果许多摄像机正在监视相同的区域,则将该区域分配给每个摄像机,以最大化得分(指标)和监控的有效性。

6.5.2 事件检测和视频质量选择算法

当本地实现缺乏时,或者商业产品的性能不足以同时保持所有必需的实时功能时,机载监视系统(ASS)的特性需要改进(图 6.4)。CV 和多媒体算法由于视频编码/压缩/选择、FN 放置的使用、网络控制、语义处理以及其他程序而出现优化。由于视觉数据是网络流量的主要来源,所以可首先应用压缩算法来最小化所需的带宽,然后利用基于权重的算法将 FANET 资源分配给要求最高的空中视频流。诸如压缩和感兴趣区域(ROI)选择等任务可以通过基于光流(OF)技术的事件检测算法或变化检测过程来实现[33-34]。OF 的使用可能会受到无人机运动的影响。然而,从帧中减去平均运动向量。无人机的运动不会干扰必要的计算事件。

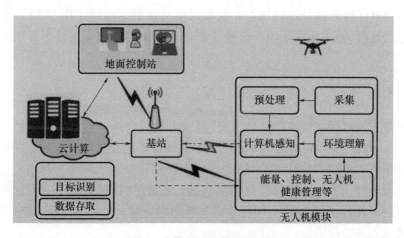

图 6.4　简化的机载监视系统

由于无人机内存受限,在运行过程中保持视频质量最高的效率是低下的。此外,当内存变满时,典型的 ASS 会覆盖最早的数据而不关心它们的内容。尽管如此,大多数捕获的图像可能包含一个事件,因此智能信息存储可以显著提高依赖于场景中事件检测的内存使用率,并使用事件指标来积累数据。考虑到有限的机载计算能力,事件检测程序可以使用计算成本低廉的 OF[33-34]。在整个预处理阶段,该过程可以通过从所有向量中减去平均值向量来归一化每个特征向量,以补偿 FN 位移,因为水平位移、倾斜或跨度将迫使所有像素假定相同的方向向量。除非 FN 接收到另一个目标区作为一个不同规模的任务,否则垂直位移并不常见。随后,模块验证 OF 值以了解是否发生任何事件。如果预定义的值 OF 值大于建立的阈值,则该元素认为该帧对应于捕获的事件。根据事件检测结果[35-38],该算法计算一些指标来评估帧的视觉质量指标并用于决策。当最近帧中没有检测到事件时,视频质量指标值逐渐减小;如果检测到一个事件,那么质量指标值将在更大程度上增加,以快速响应事件。

6.5.3 机载视频管理(无人机)

由于有限的资源:板载内存、计算能力和无线网络带宽,机载视频管理是必不可少的。当有足够的空间时,就可以存储视频。尽管如此,由于资源有限,当需要更多空间时,必须删除不太重要的视频。首先,每次存储达到其最大容量的百分比时,都需要清除一小部分存储。如果视频的事件分数大于视频候选人淘汰的平均分数或其他阈值,则将该视频标记为显著且完整;否则,删除视频。如果达到最大存储量,则无人机返回基地。由于板载计算能力也受到限制,帧间编码框架可能不可行。然而,无人机监视能力需要实时视频流。因此,我们使用了一种帧级压缩方案,一种基于 Web 的运动 JPEG(MJPEG)框架[39-41]。每一帧都可以通过事件检测得分进行压缩。视频采集一次只能通过使用 MJPEG 服务器完成一个过程,创建捕获子流。这些子流可以传输给访问流或其保存版本的操作符[42-43]。

6.5.4 舰队平台视频速率适配

即使无人机没有检测到异常情况,操作人员或管理员也应该能够检查该区域[44-45]。因此,无人机应该能够根据用户或管理员的要求提供视频流,然而,由于视频字节率很高,可以采用视频速率自适应来完全利用网络。每当发生视频流请求服务时,ASS 向 GS 报告视频流数据。因此,可以得到一个完整的网络拓扑,包括实时的视频流路由。根据数据和链路的位置,可以从一段距离估计每个链路的无人机容量[27]。

6.5.5 飞行节点协调

地面站(GS)通过向每个 FN 一致地发送全球位置请求来执行无人机群的协调[46-48]。请求-应答技术取代了简单的连续更新,使 FN 和 GS 都能识别断开的连接。FN 管理的主要目标是通过合适的位置分配 FN。通过使用布局算法,地面导航系统将每个无人机引导到它必须到达的地方。FN 的识别可以通过索引和 IP 进行。UAV 索引类似于数组列表元素的索引,每次成员关系变化时都会发生变化。当作业优先级非常重要,并且高优先级任务必须不中断时,索引方法才会非常有用。即使在 FANET 失效的情况下,FN 的 IP 也不受影响。当一架无人机试图加入 GS 时,只要连接中断,它就会返回基地。基于 IP 的标识将 FN 转回前面的操作区域(图6.5)。

图 6.5　无人机数据采集

6.5.6　数据收集和显示

每个 FN 都有一个 Web 服务器,以便外部访问其数据。这个 Web 服务器可以处理视频流和图像查询,以便可以频繁更新图像视图。服务器收集图像/视频并将它们发送到网络服务器[48-51]。无人机需要用一个专用网络连接,以便 GS 下载可视化数据,并从本地无人机的存储中获取数据并显示出来,以授予任何 GS 运营商查看图像的访问权限。服务器可以通过网页显示无人机的位置和它们的图像/视频。

6.5.7　软件定义的网络

软件定义网络(SDN)是一种网络架构,它提供了控制和数据之间的分离,它通过集中式网络可编程性提供控制和数据之间的分离(图 6.6)。SDN 由于其控制器而提高了性能,其中统一的 SDN 交换机取代了原来的网络交换机[52-56]。因此,SDN 简化了新应用程序和服务的安装和监督。此外,FANET 管理、可编程性以及重新配置都可以大大简化。FANET 可以使用 SDN 来解决其环境的缺点和性能问题,如动态和即时的拓扑修改,无人机退役时 FN 之间的链接,SDN 也可以便于处理网络管理的复杂性。OpenFlow 是一种 SDN 协议,它分离了支持 FANET 的网络控制和数据功能。SDN 控制可以是集中式、分散式(SDN 控制器分散在所有 FN 中)或在每个 FN 上本地执行的发送数据包的处理控制的混合,并且控制流量也存在于集中式 SDN 控制器和所有其他 SDN 组件。控制器从 OpenFlow 交换机收集网络统计信息和最新 FN 网络拓扑信息。因此,必须保持 SDN 控制器与无人机节点的连接。集中式 SDN 控制器具有全局网络视图。SDN 平台的交换机除了支持 FANET 和 SDN 控制之间的通信的协议外,还包括软件流程表。然后,控制器定义路径并将其发送到 FANET 的元素,在那里传输数据包。

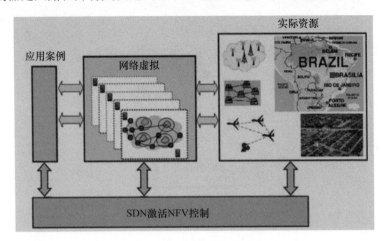

图 6.6　软件定义的网络

6.5.8　网络功能虚拟化

网络功能虚拟化(NFV)是一种原创的网络架构理念,可实现网络基础设施的虚拟化[52-56](图 6.6)。更明确地说,通过利用当前的 IT 虚拟化技术,通过 NFV 改进了网络功能配置方法。因此,虚拟机可以替代网络设备、硬件和基本功能。NFV 可以在 FANET 中提供编程功能,降低网络管理的复杂性。无人机作为一种空中基站使用多层 FN – cell 网络,FN – cell 管理结构可以利用 NFV 和 SDN 模型平衡地面网络。视频监控平台即服务(VMPaaS)可以与无人机群一起组成一个 FANET。该平台采用了最近的 NFV 和 SDN 范式。由于 NFV 互连和控制的复杂性,SDN 可以与 NFV 结合。

6.5.9　数据收集与能量收集

FANET 可以在无线网络中辅助能量收集(EH),这是由于它们的特性与固定陆地基站相比具有优势[53]。一个 FN 可以是一个飞行基站(FBS)按需流动网络服务访问,也可以为地面设备安排无线充电。FBS 可以获取数据和为耗尽的地面物联网设备进行充电。具有 EH 模块和无线充电功能的无人机可以延长网络寿命。尽管如此,FN 可以成为有需要的物联网设备的潜在能源。

相反,FBS 可以从物联网设备获取数据,前提是它们拥有足够的能量来广播它们的数据包,而 FN 同时访问发送到物联网设备/由物联网设备发送的 EH – RF 信号。无人机站由于其能量限制,覆盖时间受限。因此,必须对地面充电和信息收集时间之间的有效折中进行一些调查。还必须考虑在最小间隔内完成无人机轨迹的控制和优化方案,同时最小化分配的能量和/或增加网络寿命。大量的数值结果和模拟证实了系统在不同场景下的行为方式,并且可以研究多个指标。

物联网模式允许在连接真实世界和虚拟世界的同时收集、存储、传输和处理数据。此外,对更高的数据速率和更高的 QoS 的需求需要不断地开发。一个策略性的特点是让设备在提高能量的同时充分利用能源。能量约束使得有效的自组织至关重要,因此处理物联网设备有效消耗能量的方式和程度,从而延长其电池寿命是必不可少的。持续或按需能源供应的现行设计是最低限度的,需要工作的问题仍然悬而未决。一个重要的问题是通过什么方式对物联网设备进行无线和远程充电,因此,FBS 可以用于地面物联网设备的下行链路,同时保持其可访问性和基本流程。

通过精确的模型,无人机可以在互联物联网生态系统设备中发挥重要作用,这些模型能够捕捉到 ROI 中对象的行为。尽管如此,无人机仍具有丰富的新功能和灵活的功能,如高移动性、无处不在的网络访问以及对 GS 运营商和物联网设备的覆盖。它们在整个运行时的可重编程性、它们通过即插即用部署捕获细节和获得新度量的方式是极其重要的。此外,物联网设备的能量限制需要转换为可接受的阈值。降低能量吸收,延长网络寿命是影响超密集物联网的一个非常棘手的课题,它具有巨大的应用前景和以电池为主要能源的空间分布节点。因此,定期为大型节点充电或定期更换电池既昂贵又不方便。一个用于无线 EH 和网络的无人机站便可以解决这个问题。

为扩展 FN 的自主性和随时随地实现无线通信覆盖所做的重大努力表明,在从无人机到地面物联网设备的上行链路信息收集过程中须控制能量使用。传感器唤醒时间表和 FN 轨迹的相互优化以完成任务(具有可靠和节能的数据采集)可以更好地组织无人机的能源,无人机最小单元(USC)网络的最优组织可以根据子模块博弈的观点来完成。主要目标是在扩大能源效率的同时,满足合理的网络可用性和足够的相遇率。

此外,能源效率和吞吐量可以采用大型天线框架作为最佳形式,以减少基站的碳排放量。解决三维 FN 放置问题的最好的方法是通过减少要定位的移动基站(MBS)的数量,而每个 GS 必须保持在至少一个 MBS 通信范围内。当假设空对地链路中只能发生视距(LoS)时,可能会产生路径损耗,从而实现具有最佳网络性能的 FN 基站的最佳尺寸。此外,还可以将低空平台与 USC 的下行覆盖性能结合起来,以实现 USC 的高效部署。利用 FN 作为基站有助于最大限度地提高覆盖概率,而覆盖概率取决于网络性能无人机高度和发射功率。

在给容易受到恶意攻击的无人驾驶飞机提供服务的同时,必须加强和平衡 FN 对 CPS 攻击的安全性。到目前为止,这些模型提供了联合网络访问和无线能量传输,其中EH能力不足以覆盖/解决FN物联网区域。例如,如果FN通过射频传输能量到一个设备上,但是受到的能量有限,并不是由一个实质性的多输入多输出(MIMO)数据中心直接支持,则将会遵循几个安全程序。当FN负责无线功率传输时,考虑到其移动性,可以发现最优路径,最大限度地提高输送到地面接收器的总能量。

物联网要求EH解决方案提供自主自供电系统,以从外部环境能源(如太阳能、热能和风能)生产能源。在基于无线电的网络中,通过射频源获取能量是更好的选择,因为它易于实现且增益较低。配备EH装置的车站可以将接收到的射频信号转换成能量,以使电池设备恢复活力。RF-EH可以延长传感器和执行器的使用寿命,同时降低电池更换成本。

一个同样需要注意的问题是,从物联网设备收集的数据接入点(DAP)和向设备传输能量的能量接入点之间的信息不对称。一个最佳的信息收集性能可以依赖于缓存和阈值调节程序,以减少传输到一个标记传感器的请求数。然而,在某些框架中,D2D网络中的EH被认为是机器型通信(MTC)和蜂窝网络的基础。

同样地,D2D用户花费在MTC业务流量和能量消耗其可用蜂窝频谱上的最佳时间必须得到协调。此外,安全性是EH应用程序中一个令人注意的话题。RF-EH的学习结构允许混合接入点在存在对抗性学习问题时确定足够的功耗。毫无疑问,卸载计算可以通过FN友好的物联网平台帮助检测和识别人群中的不信任个体,从而改善无人机的反应性。

6.6 小结

FANET的特征是具有快速链接更改的流体拓扑,其中每个无人机是具有三维移动性和节点之间快速链路质量变化的FN。因此,通信协议的设计面临不同的挑战,需要创新的形式来实现通信的前提条件。

FANET的瓶颈是:① UAV-GS、UAV-卫星和U2U之间的无线通信;② 当频繁发生连接中断时,每架无人机必须有路由、协调、协作和通信协议;③ 允许高性能流体网络拓扑;④ 无人机动力资源有限。因此,FANET需要一些具有新颖网络范式的专业硬件。

无人机功率限制,限制了无人机的计算能力、通信和续航能力。采用低功耗和有损网络(LLT)方法的能源感知无人机机队部署是处理此问题的一种选择。

鉴于FANET的独特特点,SDN和NFV是解决复杂资源管理的一种比较好的方法。

FANET 还将与呈指数增长的无线视觉传感器和驱动网络进行交互。因此，必须认真地建立协议和策略以便进行这种扩展[54-57]。

参考文献

[1] V. V. Estrela, O. Saotome, H. J. Loschi, et al., "Emergency Response Cyber – Physical Framework for Landslide Avoidance with Sustainable Electronics," *Technologies*, vol. 6, p. 42, 2018. doi:10.3390/technologies6020042.

[2] O. K. Sahingoz, "Networking Models in Flying Ad – Hoc Networks (FANETs): Concepts and Challenges," *J. Intell. Robot. Syst. Theory Appl.*, vol. 74, no. 1 – 2, pp. 513 – 527, 2014.

[3] O. K. Sahingoz, "Mobile Networking with UAVs: Opportunities and Challenges," *2013 Int. Conf. Unmanned Aircr. Syst. ICUAS 2013 – Conf. Proc.*, Atlanta, GA, USA, pp. 933 – 941, 2013.

[4] I. Bekmezci, O. K. Sahingoz, and Ş. Temel, "Flying Ad – Hoc Networks (FANETs): A Survey," *Ad Hoc Networks*, vol. 11, no. 3, pp. 1254 – 1270, 2013.

[5] A. Guillen – Perez and M. – D. Cano, "Flying Ad Hoc Networks: A New Domain for Network Communications," *Sensors*, vol. 18, p. 3571, 2018.

[6] K. Palan and P. Sharma, "FANET Communication Protocols: A Survey," *Ijcsc*, vol. 7, no. 1, pp. 219 – 223, 2015.

[7] A. I. Alshabtat and L. Dong, "Adaptive MAC Protocol for UAV Communication Networks Using Directional Antennas," *2010 Int. Conf. Networking, Sens. Control*, Chicago, IL, USA, 2010, pp. 598 – 603.

[8] Y. Cai, F. R. R. Yu, J. Li, Y. Zhou, and L. Lamont, "MAC Performance Improvement in UAV Ad – Hoc Networks with Full – Duplex Radios and Multi – Packet Reception Capability," Ottawa, ON, Canada, pp. 523 – 527, 2012.

[9] A. Alshabtat, L. Dong, J. Li, and F. Yang, "Low Latency Routing Algorithm for Unmanned Aerial Vehicles Ad – Hoc Networks," *Int. J. Electr. Comput. Energ. Electron. Commun. Eng.*, vol. 6, no. 1, pp. 48 – 54, 2010.

[10] S. Defense, S. Symposium, J. H. Forsmann, R. E. Hiromoto, and J. Svoboda, "A TimeSlotted On – Demand Routing Protocol for Mobile Ad Hoc Unmanned Vehicle Systems A Time – Slotted On – Demand Routing Protocol for Mobile Ad Hoc Unmanned Vehicle Systems," 2007.

[11] R. Bilal and B. M. Khan, "Analysis of Mobility Models and Routing Schemes for Flying Adhoc Networks (FANETS)," *Int. J. Appl. Eng. Res.*, vol. 12, no. 12, pp. 3263 – 3269, 2017.

[12] C. Zang and S. Zang, "Mobility Prediction Clustering Algorithm for UAV Networking," *2011 IEEE GLOBECOM Work. GC Wkshps*, Houston, TX, USA, 2011, no. 3, pp. 1158 – 1161, 2011.

[13] E. Yanmaz, C. Costanzo, C. Bettstetter, and W. Elmenreich, "A Discrete Stochastic Process for Coverage Analysis of Autonomous UAV Networks," *2010 IEEE Globecom Work. GC'10*,

pp. 1777 - 1782,2010.

[14] E. W. Frew and T. X. Brown, "Networking Issues for Small Unmanned Aircraft Systems," *J. Intell. Robot. Syst.* , vol. 54, no. 1 - 3, pp. 21 - 37, 2009.

[15] H. Zhai, Y. Kwon, and Y. Fang, "Performance Analysis of IEEE 802.11 MAC Protocols in Wireless LANs," *Wirel. Commun. Mob. Comput.* , vol. 4, no. 8, pp. 917 - 931, 2004.

[16] N. Walravens and P. Ballon, "Platform Business Models for Smart Cities: From Control and Value to Governance and Public Value," *Commun. Mag.* IEEE, no. June, pp. 72 - 79, 2013.

[17] A. Purohit, F. Mokaya, and P. Zhang, "Collaborative Indoor Sensing with the Sensorfly Aerial Sensor Network," *Proc. 11th Int. Conf. Inf. Process. Sens. Networks - IPSN'12*, p. 145, 2012.

[18] M. Quaritsch, K. Kruggl, D. Wischounig - Strucl, et al., "Networked UAVs as Aerial Sensor Network for Disaster Management Applications," *Elektrotechnik und Informationstechnik*, vol. 127, no. 3, pp. 56 - 63, 2010.

[19] D. Wu, Y. T. Hou, W. Zhu, Y. - Q. Zhang, and J. M. Peha, "Streaming Video over the Internet: Approaches and Directions," *IEEE Trans. Circuits Syst. Video Technol.* , vol. 11, no. 3, pp. 282 - 300, 2001.

[20] E. Setton, T. Yoo, X. Zhu, A. Goldsmith, and B. Girod, "Cross - Layer Design of Ad Hoc Networks for Real - Time Video Streaming," *IEEE Wirel. Commun.* , vol. 12, no. 4, pp. 59 - 65, 2005.

[21] R. Muzaffar, "Routing and Video Streaming in Drone Networks," D. Sc. dissertation, Queen Mary University of London, London, 2017.

[22] Y. M. Chen, L. Dong, and J. S. Oh, "Real - Time Video Relay for UAV Traffic Surveillance Systems through Available Communication Networks," *2007 IEEE Wireless Communications and Networking Conference*, Kowloon, China, pp. 2608 - 2612, 2007. DOI: 10.1109/WC-NC.2007.485.

[23] C. C. Baseca, J. R. Diaz, and J. Lloret, "Communication Ad Hoc Protocol for Intelligent Video Sensing Using AR Drones," *Proc. - IEEE 9th Int. Conf. Mob. Ad - Hoc Sens. Networks, MSN 2013*, Dalian, China, pp. 449 - 453, 2013.

[24] J. Montgomery, S. I. Roumeliotis, A. Johnson, and L. Matthies, "The Jet Propulsion Laboratory Autonomous Helicopter Testbed: A Platform for Planetary Exploration Technology Research and Development," *J. F. Robot.* , vol. 23, no. 1, pp. 245 - 267, 2006.

[25] M. Erdelj, E. Natalizio, K. R. Chowdhury, and I. F. Akyildiz, "Help from the Sky: Leveraging UAVs for Disaster Management." *IEEE Pervasive Computing*, 2017, 16, 24 - 32.

[26] A. Detti, P. Loreti, N. Blefari - Melazzi, and F. Fedi, "Streaming H.264 Scalable Video over Data Distribution Service in a Wireless Environment," *Proc. IEEE WoWMoM*, Montreal, QC, Canada, 2010.

[27] J. Jung, S. Yoo, W. G. La, D. R. Lee, M. Bae, and H. Kim, "AVSS: Airborne Video Surveillance System," *Sensors*, 18(6): 1939, 2018.

[28] T. Ahn, J. Seok, I. Lee, and J. Han, "Reliable Flying IoT Networks for UAV Disaster Rescue Operations." *Mobile Information Systems*, 2018, 2572460: 1 - 2572460: 12. https://doi.org/10.1155/2018/2572460.

[29] R. Fan, J. Cui, S. Jin, K. Yang, and J. An, "Optimal Node Placement and Resource Allocation for UAV Relaying Network," *IEEE Communications Letters*, vol. 22, pp. 808 – 811, 2018.

[30] C. C. Lai, C. – T, Chen, and L. – C. Wang, "On – Demand Density – Aware UAV Base Station 3D Placement for Arbitrarily Distributed Users With Guaranteed Data Rates," *IEEE Wireless Communications Letters*, vol. 8, pp. 913 – 916, 2019.

[31] D. J. Hemanth and V. V. Estrela, "Deep Learning for Image Processing Applications," *Adv. Par. Comp.*, IOS Press, Netherlands. ISBN978 – 1 – 61499 – 821 – 1 (print) 978 – 1 – 61499 – 822 – 8 (online) 2017.

[32] N. Razmjooy and V. V. Estrela. "Applications of Image Processing and Soft Computing Systems in Agriculture," IGI Global, Hershey, PA, USA, pp. 1 – 337, 2019. doi: 10.4018/978 – 1 – 5225 – 8027 – 0.

[33] H. R. Marins and V. V. Estrela, "On the Use of Motion Vectors for 2D and 3D Error Concealment in H.264/AVC Video," *Feature Detectors and Motion Detection in Video Processing*. IGI Global, Hershey, PA, USA, pp. 164 – 186, 2017. doi: 10.4018/978 – 1 – 5225 – 1025 – 3. ch008.

[34] V. V. Estrela and A. M. Coelho. "State – of – the – Art Motion Estimation in the Context of 3D TV." In *Multimedia Networking and Coding*, ed. R. A. Farrugia and C. J. Debono, IGI Global, Hershey, PA, USA, pp. 148 – 173, 2013. doi: 10.4018/978 – 1 – 4666 – 2660 – 7. ch006.

[35] V. Chandola, A. Banerjee, and V. Kumar, "Anomaly Detection: A Survey," *ACM Comput. Surv.*, 41: 15:1 – 15:58, 2009.

[36] A. Farzindar and W. Khreich, "A Survey of Techniques for Event Detection in Twitter," *Computational Intelligence*, vol. 31, pp. 132 – 164, 2015.

[37] R. Schuster, R. Mörzinger, W. Haas, H. Grabner, and L. Van Gool. "Real – Time Detection of Unusual Regions in Image Streams," *Proc. MM'10: Proceedings of the 18th ACM international conference on Multimedia* 2010, Seoul, Korea, 1, pp. 1307 – 1310, 2010.

[38] K. Zhang, T. Manning, S. Wu, et al. "Capturing the Signature of Severe Weather Events in Australia Using GPS Measurements," *IEEE Journal of Selected Topics in Applied Earth Observations and Remote Sensing*, vol. 8, pp. 1839 – 1847, 2015.

[39] T. Richter, A. Artusi, and T. Ebrahimi, "JPEG XT: A New Family of JPEG Backward – Compatible Standards," *IEEE MultiMedia*, vol. 23, pp. 80 – 88, 2016.

[40] J. Theytaz, L. Yuan, D. Mcnally, and T. Ebrahimi, "Towards an Animated JPEG," *Proceedings SPIE Optical Engineering + Applications*, *Volume 9971, Applications of Digital Image Processing XXXIX*; 99711X, San Diego, California, USA, 2016. DOI: 10.1117/12.2240283.

[41] S. Mahmoud and Nader Mohamed. "Collaborative UAVs Cloud," 2014 *International Conference on Unmanned Aircraft Systems (ICUAS)*, Orlando, FL, USA, pp. 365 – 373, 2014.

[42] M. Bae and K. Hwangnam, "Authentication and Delegation for Operating a Multi – Drone System," *Sensors*, 19(9):2066, 2019.

[43] M. Bae, S. Yoo, J. Jung, et al., "Devising Mobile Sensing and Actuation Infrastructure with Drones," *Sensors*, 2018.

[44] X. Wang, X. A. Chowdhery, and M. Chiang, "SkyEyes: Adaptive Video Streaming from UA-

Vs," Proc. 3rd Workshop on Hot Topics in Wireless HotWireless'16, HotWireless@ MobiCom, New York, NY, USA, pp. 2 – 6, 2016. DOI: 10. 1145/2980115. 2980119

[45] A. Rozantsev, M. Salzmann, and P. Fua, "Beyond Sharing Weights for Deep Domain Adaptation," *IEEE Trans. on Pattern Analysis and Machine Intelligence*, vol. 41, pp. 801 – 814, 2018.

[46] M. Chen, F. Dai, H. Wang, and L. Lei, "DFM: A Distributed Flocking Model for UAV Swarm Networks," *IEEE Access*, vol. 6, pp. 69141 – 69150, 2018.

[47] D. Albani, T. Manoni, D. Nardi, and V. Trianni, "Dynamic UAV Swarm Deployment for Non – Uniform Coverage," *Proc. 2018 International Conference on Autonomous Agents and Multiagent Systems, AAMAS*, Stockholm, Sweden, 2018.

[48] J. Zhang, Y. Zeng, and R. Zhang, "UAV – Enabled Radio Access Network: Multi – Mode Communication and Trajectory Design," *IEEE Transactions on Signal Processing*, vol. 66, pp. 5269 – 5284, 2018.

[49] J. Sun, B. Li, Y. Jiang, and C. – Y. Wen. "A Camera – Based Target Detection and Positioning UAV System for Search and Rescue (SAR) Purposes," *Sensors*, 16(11), 1778, 2016.

[50] Z. Yang, R. Wang, Q. Yu, et al., "Analytical Characterization of Licklider Transmission Protocol (LTP) in Cislunar Communications," *IEEE Transactions on Aerospace and Electronic Systems*, vol. 50, no. 3, pp. 2019 – 2031, 2014.

[51] L. Gupta, R. Jain, and G. Vaszkun, "Survey of Important Issues in UAV Communication Networks," *IEEE Communications Surveys & Tutorials*, vol. 18, no. 2, pp. 1123 – 1152, 2016.

[52] H. Shakhatreh, A. H. Sawalmeh, A. Al – Fuqaha, *et al.* "Unmanned Aerial Vehicles (UAVs): A Survey on Civil Applications and Key Research Challenges," *IEEE Access*, vol. 7, pp. 48572 – 48634, 2019. doi: 10. 1109/ACCESS. 2019. 2909530.

[53] S. Arabi, E. Sabir, H. Elbiaze, and M. Sadik, "Data Gathering and Energy Transfer Dilemma in UAV – Assisted Flying Access Network for IoT," *Sensors*, vol. 18, pp. 1519, 2018. doi: 10. 3390/s18051519.

[54] V. V. Estrela, A. C. B. Monteiro, R. P. França, Y. Iano, A. Khelassi, and N. Razmjooy, "Health 4. 0 as an Application of Industry 4. 0 in Healthcare Services and Management," *Medical Technologies Journal*, vol. 2, no. 4, pp. 262 – 276, 2019. doi: 10. 26415/2572 – 004X – vol2iss1p262 – 276.

[55] V. V. Estrela, A. Khelassi, A. C. B. Monteiro, *et al.* "Why software – defined radio (SDR) matters in healthcare?". *Medical Technologies Journal*, vol. 3, no. 3, Oct. 2019, pp. 421 – 9, DOI: 10. 26415/2572 – 004X – vol3iss3p421 – 429.

[56] D. Kreutz, F. M. V. Ramos, P. Verissimo, *et al.* "Software – Defined networking: A comprehensive survey," *Proceedings of the IEEE*, vol. 103, pp. 14 – 76, 2014.

[57] S. Sudevalayam, and P. Kulkarni, "Energy harvesting sensor nodes: Survey and implications." IEEE Communications Surveys & Tutorials, 13, pp. 443 – 461.

第7章　无人机多光谱与高光谱成像：现状与展望

过去数十年来，成像传感设备在监视、监控、跟踪和摧毁空间物体等各种任务中表现得越来越先进。目前，UAV 已非常普及，它可以获得全方位视野，甚至能在极低的距地目标高度上执行任务。与其他远程任务相比，UAV 的开发成本极低，因此具有极高的效费比。本章旨在详细介绍两种不同 UAV 遥感(RS)技术：① 多光谱成像(MSI)；② 高光谱成像(HSI)，解释了所观测到的潜在自然现象的空间和光谱特征。

7.1　引言

科技的飞速发展推动了大型工业机械乃至汽车工业的自动化。自动驾驶汽车上路已不是遥不可及的梦想，一些领先企业，如美国谷歌公司、优步科技公司和特斯拉汽车，正在努力在每条道路上实现这一梦想[1,2]。采用 RS，如卫星成像技术，由于增加了光谱和空间方面更多的详细信息，从而具有以更高分辨率观测景区的巨大优势。在 UAV 采用这些成像设备后，可进一步利用图像改善各区域的空间目标定位能力和关注区域的变化检测能力。

MSI 传感器有助于景区分析，其光谱分辨率仅限于可见光和红外(IR)光谱区域的几个波段，而 HSI 可包含甚至上述同一光谱区域内观测到数百个光谱通道。对于 HSI 传感器，存在推扫和快照两种配置模式，两者不同之处在于是同时覆盖空间区域，还是通过一次扫描一个条带区域来覆盖景区。采用这些成像设备的优势在于可观测到待观测复杂空间目标的高分辨率光谱特征。整个 UAV 系统作为一种信息－物理系统(CPS)(UAV－CPS)，在国防和监视任务专业救生行动中，可能更具瞬时图像识别优势，更高水平的光谱识别能力在这些任务中作用明显。这为采用 MSI 和 HIS 传感器进行 UAV 成像铺平了道路。

在卫星成像中,空间分辨率决定了可获得更优的空间物体清晰度。通过这些传感设备捕获的图像只是空间物体以数字的形式记录下来的反射率,但观测高度越高,成像就会越复杂。不过,成像传感设备经过不断发展,已使监测农田、单株树高和植被健康变得越来越容易。在发生自然灾害时,通过在该区域启用UAV成像,可即时获取并充分估计环境的实时变化和遭受破坏的情况[3-4]。通过地面和卫星传感器成像的方式,很难获得表层下空间物体的更具体信息。因此,在最近的空中监测或土地利用/变化检测应用中,UAV成像的使用发生了巨大的爆炸式增长。相关UAV成像在导航通信、交通监控和许多其他监视应用中通过提升人体舒适度,降低了体力劳动强度[5]。

无人空中平台实际上已成为致命武器和非致命武器的新兴选择,并使作战发生了巨大的变革。UAV的最大优势是可在战场上自动驾驶飞行,从而防止了人员伤亡。UAV可用于空中监视等相关活动。然而,对于一般用户来说,全自动化的UAV仍然是一种先进的高成本支出工具。通常从这些传感设备获取的图像包含不同光谱带(对应各自波长)的详细信息,如图7.1所示。

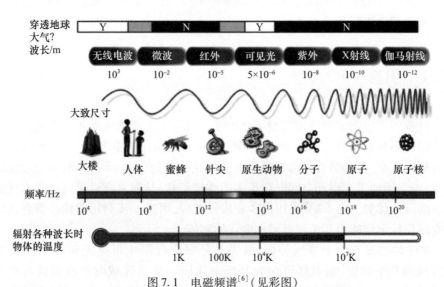

图 7.1 电磁频谱[6](见彩图)

可分析光谱区域的反射,通过传感来绘制健康和患病植被,如图7.2所示。通常,植物中绿色和近红外(NIR)反射率占优;这就是采用这种波段组合来推导植被指数的原因所在。如果NIR区域反射率更高,则叶子是健康的,否则(树叶凋落/患病)反射率会降低,这说明了在对目标场景进行成像时观察不同频段光谱的重要性。

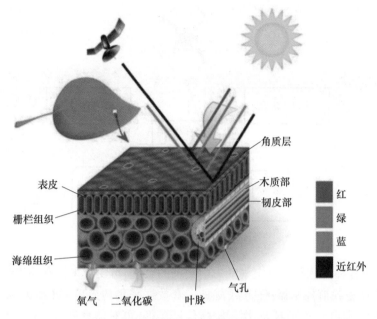

图 7.2　基于光谱分析的植被图（见彩图）

本章详细介绍了以下概念。
(1) UAV 的组成和运行,描绘了机上搭载的 MSI 和 HIS 设备有限的能力[7-8]。
(2) UAV 所采集的空间数据的光谱与空间分辨率的重要性。
(3) 卫星成像设备与 UAV 的 MSI 和 HIS 对比。
(4) UAV 机载采集和近距地面采集支持各种研究应用,甚至可用于监控路上人员的活动及类似的交通活动[9-10]。
(5) MSI 和 HIS 采用的图像维度和数据缩减技术[11-12]。

7.2　UAV 成像体系架构和组成

一般来说,UAV 由飞行控制系统的自动化嵌入式系统进行驾驶,其系统内含各种传感器,如加速度计、陀螺仪、全球定位系统(GPS)、压力传感器等用来执行飞行计划中的 UAV 任务。每架 UAV 由 5 个模块组成,以执行航空测量任务[13-14]。图 7.3 说明了 UAV 的总体架构:① 飞行计算机;② 任务和载荷控制;③ 通信子系统;④ 数字相机和传感器设备;⑤ 陀螺稳定观测平台。

UAV 的外部组件简单、轻便且稳定,但容纳传感设备和复杂技术仪器的空间有限。

(1) 飞行计算机:该组件是 UAV 的中央控制单元,一种通过加速度计、磁强

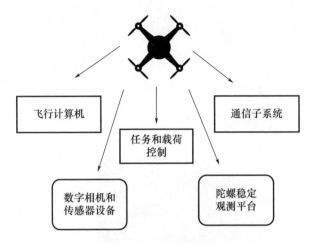

图 7.3 UAV 的总体架构

计、GPS 等各种传感器采集空气动力学数据的系统,可实现飞机自动飞行。

(2)任务载荷控制器:任务载荷是一套数据采集传感器,如 RGB 摄像头、红外、热、MSI 和 HSI 传感器等,用于实施信息采集(可对信息进行一定程度的即时处理,或采用后处理)。此外,任务控制器还可根据飞行计划控制传感器载荷。

(3)通信基础设施:整个系统的主干,所有模块都是通过混合通信机制实施同步控制的,从而将确保 UAV 和基站之间的可靠链接。尽管 UAV 系统是高度自动化的,但仍强制要求地面站执行适度的人工控制。飞行控制计算机可根据飞行计划执行操控,但无法在自动处理中执行超出设定的行为和反应。类似地,这些载荷也主要通过半自动化方式实现远程操控。

图 7.4 详细说明了使用 UAV 进行测量飞行活动的信息流。首先,地面站针

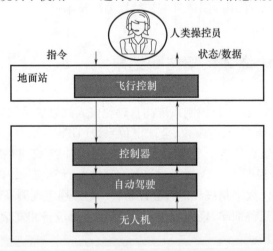

图 7.4 UAV 工作流程图

对测量活动发出飞行命令与飞行前计划。控制器用作控制和协调成像及飞行的中央单元。

总的来说,当前 UAV 的自动驾驶功能已非常先进,在出现电池电量不足或任何此类硬件故障的情况下,安全返回模式会使 UAV 安全返航,甚至可借助归一化差异植被指数(NDVI)处理相机进行瞬时图像处理。标准的 RGB 摄像头只能进行可见光成像,而采用 NDVI 处理的改进型相机可进行 NIR 成像。

7.2.1 UAV 的未来应用领域

UAV 在多个不同领域的成功推动了全球范围内大量研究工作的开展,以突破限制利用 UAV。相关研究重点关注以下领域[15-17]。

(1)复杂航空目标检测。
(2)陆、海、空三栖 UAV。
(3)各个运行域的自动定位。
(4)UAV 飞行降噪。
(5)坚固防水外壳,适应水下活动。
(6)采用 GPS 自动导航进行海岸搜索作业。
(7)目标识别的智能化自学习能力。
(8)提高载荷应力水平,便于装载执行水下和空中任务的传感器和设备。

UAV-CPS 在能量和载荷方面面临着严格的限制,通过云计算、FANET 节点的智能化应用和更优的图像表示方法可在一定程度上缓解这些限制。随着语义和元数据投入应用,将减少机载存储量[18-20]。

7.3 多光谱与高光谱成像仪

自 1960 年首颗遥感卫星(TIROS-1)发射以来[21],用于大气、土地覆盖、自然植被和沿海地区等各种研究的遥感器的发射数量逐年提高。由于空间分辨率不同,这些遥感器产生了精度水平各不相同的各类图像。采用图像数字化像素[22]对所覆盖表面区域进行测量。对于 UAV 系统,这些 MSI 和 HSI 传感器有助于有效实现土地覆盖和农业状况绘图,如图 7.5 所示。

7.3.1 多光谱成像

MSI 传感器感知不同波长的光谱,而卫星轨道高度是决定其空间分辨率精度的主要因素。MSI 由 3 个或 4 个宽波段以及 1 个全色数据波段组成,这些波段不

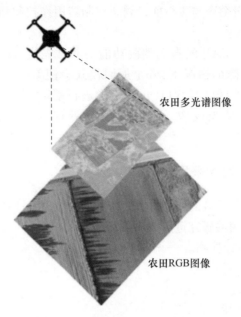

图 7.5 多光谱成像(见彩图)

仅包括人眼能够感知的可见光谱(400~700 nm)区域,还进一步延伸至各种红外波段,如近红外、中红外、热红外和其他紫外波段[23]。

7.3.1.1 低分辨率成像

30~1000m 的低分辨率 MSI 有助于农业监测和其他能源应用。这里,空间分辨率取决于像素数,即像素数越多,图像清晰度越高[24]。低分辨率 MSI 被广泛应用于大面积土地覆盖分析,图像融合或配准技术加上超分辨率技术可提高 MSI 图像的质量。

7.3.1.2 高分辨率成像

高分辨率 MSI 的对地空间分辨率从 4m 到数厘米不等,主要用于监视用途[25]。更高分辨率的图像可用于小片土地区域识别与绘制、后勤规划、城市发展、居住地制图等[26]。MSI 的典型限制在于对空间目标粗略的分辨率和极低的辨识能力。通过使用简单的特征集(如颜色、纹理和形状等)来区分复杂的空间目标,只会带来不太准确的结果[27]。

7.3.2 高光谱成像

HSI 传感器包含数千个光谱波段,这些波段极窄(10~20nm),也就是说,每个数字化像素具有许多相邻光谱波段的光强度[28]。因此,图像的每个数字化像素都具有一个详细的连续光谱,可通过更准确的信息表征空间目标。

Hyperion 成像光谱仪是一种知名 HSI 传感器,其采用 30m 空间分辨率和强辐射光谱法来记录 220 个光谱带(0.4~2.5mm)[29]。图 7.6 显示了 MSI 和 HSI 在空间精度方面的变化。

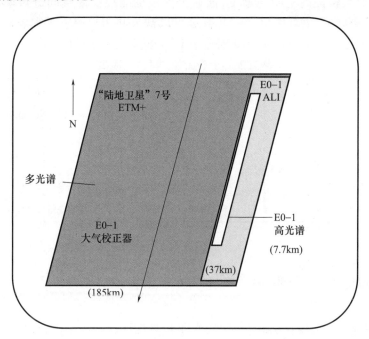

图 7.6　空间分辨率:MSI 比对 HSI[30](见彩图)

与现有的 MSI 专题制图仪器相比,HIS 取得的详细观测结果为土地覆盖分类研究提供了更多的证据。HSI 被广泛应用于各种实际应用,如矿产勘探、监测[31]、利用 AVHRR 进行土地覆盖研究[32]、植物种类监测[33]和滑坡监测[34],其他知名的 HIS 仪器还包括 PROBA、PRISMA、IMS-1 和 HySI 等。此外,HSI 数据处理的重大挑战源于特定波长内大量光谱通道细节造成的冗余信息,以及海量数据和图像校准限制。表 7.1 列出了广泛使用的 HSI 研究数据集和仪器参数。

表 7.1　高光谱数据集

编号	数据集	研究区域	传感器设备	光谱波段
1	印度松树	印第安纳州西北部	AVIRIS 传感器	224
2	帕维亚大学	意大利北部的帕维亚	ROSIS 传感器	103
3	萨利纳斯	加利福尼亚州的萨利纳斯谷地	AVIRIS 传感器	224
4	博茨瓦纳	博茨瓦那的奥卡万戈三角洲	NASA EO-1	242
5	肯尼迪航天中心	佛罗里达州	NASAAVIRIS	176

7.3.3 卫星成像与 UAV 成像

卫星成像仪是遥感技术出现以来广泛应用的空中成像方式,但采用 MSI 和 HIS 并没有大幅提高空间清晰度,商业卫星极少能提供高空间分辨率图像,而 UAV 图像更精确,更具体化。UAV 甚至可以被公共用户用于管理农田,而卫星成像更关注于国家应用。表 7.2 对比了 UAV 成像与卫星成像。

表 7.2 卫星成像与 UAV 成像

影响因子	UAV 成像	卫星成像
空间分辨率	亚厘米级(<6m)	空间清晰度相对较低 如陆地卫星为 30m
图像采集频率	每天或生长季的关键期	每月平均 4 次
天气干扰	在大雨期间无法飞行	云层覆盖和其他大气扰动
图像滤波器	窄光谱波段滤波器	固定宽带光谱波段滤波器

7.4 UAV 图像处理工作流程

机载或星载成像仪出现在气候、农业和监视等众多应用中。由于大气扰动和时间变化,这些卫星图像易受噪声的影响。可通过从被认可的数据中心获取空间数据来进行正射校正。然而,这些图像在用于分类或其他空间数据任务前,还需要对它们进行一些处理[35]。

7.4.1 节详细介绍空间数据增强和挖掘的大致处理流程。空间数据处理包括:① 大气校正;② 光谱影响;③ 降维(DR);④ 计算任务(分割、特征提取、分类或检测)。

7.4.1 大气校正

这些数字图像源于地表反射的辐射。由于大气层会发生光的散射而引起大气扰动,这些图像包含有一些噪声,可通过大气校正在一定程度上降低这些噪声。由于云层扰动会造成极少量的光谱混淆,噪声校正和补偿过程在利用空间数据开展变化检测研究中必不可少。极有效的去云算法仍然是一个开放的挑战,而阈值化或掩蔽云区的效果不佳[36]。可采用标准的大气校正方法,如暗物体扣除法(DOS1,DOS2)。除此之外,FLAASH、6S 方法也广泛用于大气校正[37]。尽管极少有地理信息系统(GIS)处理工具内置预处理功能,但形态学图像处理操作已广泛应用于空间图像的去噪。通常采用直方图均衡化方法和对比度增强方法,但由于较高的对比度,可能降低图像质量[38-39]。伽马校正也可用于增强图像,其中伽马值高于或低于 1 会导致适度的亮度增强[40]。空间图像中的降噪可采用平滑图像滤波器来进行平均或模糊处理,这可能会在一定程度上降低图像质量。

7.4.2 光谱影响映射

图 7.7 光谱反射率(SR)图显示出所有的曲线在可见光谱中更为靠近,即所有这些空间物体的反射率在红外区域都可以得到很好的区分。

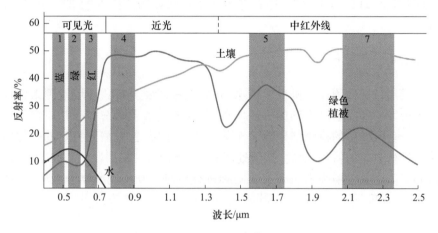

图 7.7　光谱反射率图[41](见彩图)

图 7.7 中,曲线的较低值说明这些光谱波段中光的吸收率高。例如,植被类在红波段吸收率增高会导致红外波段反射率增高。这种现象表现为红边效应,相对类似于天文观测中检测到的多普勒效应的红移效应,是观测者和波源(任何波介质)在相互接近或碰面时频率变化的观测结果[42]。

基于这些光谱波段的光谱特性,将各种光谱指标引入到空间目标监测的实践中。基于卫星图像的衍生产品,如归一化植被指数和归一化湿度指数,是最常用的作物健康监测和水域变化管理光谱指数[43-44]。类似地,各种光谱波段组合也用于了许多其他植被、气候和水质指数[45-47]。

7.4.3 降维

由于涉及各种波长的波段,因此 HSI 数据量巨大,这类数据需要通过 DR 来降低计算成本,同时克服地面真值(GT)的限制。在利用 HSI 数据开展土地覆盖绘图等应用中,光谱空间分类需要合适的 GT。因此,可采用较低维度的特征空间进行处理[48]。目前,广泛应用的 DR 技术有最小噪声分离(MNF)、独立分量分析和遗传算法(GA)等,其中 GA 是一种众所周知的特征优化技术。MNF 变换只是作为线性变换的两个级联主成分分析[49-50]。它取决于信噪比,因此不受波段范围变化(在多波段分析中不可避免)的影响。多波段数据的维度也可通过选择一个充足的数据子集来进行降低,从而不会影响空间类的稳健判别特征。GA 是这类波段选择的最佳选择之一。GA 最常见的缺点是多种群遗传算法(MGA)的早熟收敛问题。它包括两个并行的子种群,迁移策略是在合适的各代间隔内按照迁移率进行最佳 - 最差交换。MGA 方法在分类误差极小的情况下比简单遗传算法的分类效果更好。上述光谱反射率技术只是适于空间数据处理的若干示例,还有许多优化算法适用于高效的波段子集/特征选择。

7.4.4 计算任务

可采用目标检测计算机视觉模型,开展卫星图像分析式环境监测来跟踪自然灾害爆发。这能通过不同的方法实现,如非监督或监督分类方法。空间判别特征对于获取准确的检测结果起着至关重要的作用,这就需要特定的图像增强方法来进行有效的特征识别。从预处理到特征提取,可采用颜色空间转换、图像滤波和阈值函数等增强方法来突出目标进行识别。在空间数据开发中,广泛使用核机器、深度学习模型乃至贝叶斯概率分类估计等智能方法[51-52]。

7.5 空间数据的数据处理工具包

传感仪器的革新自然而然推动了开发空间数据处理软件包和工具包。数年前,基本上只有商业化的地理空间软件包才具有可视化更高空间分辨率和提取光谱信号特征进行分类等先进功能。然而,当前采用开源工具包和插件便能执行空间数据处理的完整处理周期[53],对环境监测的关注已启迪发展迅猛的研究团体和机器学习爱好者在空间数据分析上开展更高水平的研究。在使用许多开源工具包时会发现的一个重大限制,即需要将输入数据集转换为适当的扩展名(.lan,.hdr,.mat,.img)。除此之外,这些软件包极少为那种需要频繁更新的命令行 beta 版本。

表 7.3 列出了空间数据爱好者所关注和广泛使用的各种工具包和软件包,这些工具尤其适用于 UAV 图像。

第 7 章　无人机多光谱与高光谱成像：现状与展望

表 7.3　MSI 和 HIS 空间数据分析工具

编号	软件/插件	特性/用途	来源
1	Spectral Python（SPy）	用于 HSI 处理的 Python 模块	http：//www.spectralpython.net/
2	InterImage	自动图像处理功能	http：//www.lvc.ele.puc-rio.br/projects/interimage/
3	VMS 摄影测量软件	UAV 图像处理工具	http：//geomsoft.com
4	Quantum GIS	广泛使用的开放式地理信息系统，直接数据下载和半自动分类插件	https：//www.qgis.org/en/site/
5	Open Jump	大数据集处理、地形工具	http：//www.openjump.org
6	Pix4Dag	地表反射率产品和指数图	https：//sysmap.net/pix4dag
7	Sentera 公司 Agvault	电子农业：跟踪作物生长、杂草和暴风雨损失	https：//sentera.com/fieldagent-platform/
8	Sensefly 公司的 seBee	三种不同的无人机飞行与分析软件包（Airware），支持不同的应用	https：//www.sensefly.com/software/
9	谷歌地球专业版	空间数据可视化和三维地图生成	https：//www.google.com/earth/download/gep/agree.html
10	谷歌地球引擎	地理空间可视化；高维历史数据的大规模科学分析	https：//earthengine.google.com/
11	Orthomosaic Blend Code	正射镶嵌的处理管道	http：//build-mosaic.sf.net
12	Multispec	MSI 和 HSI 数据分析	https：//engineering.purdue.edu/~biehl/MultiSpec/download_win.html
13	开放数据工具包-ODK	支持从移动设备到服务器的数据收集	https：//opendatakit.org/
14	PostgreSQL/PostGIS	对象关系数据库管理；大规模空间数据分析	https：//postgis.net/
15	GeoServer	地理空间数据管理与资源共享	http：//geoserver.org/

129

(续)

编号	软件/插件	特性/用途	来源
16	Drone Deploy—三维模型生成工具	三维变换	https://www.dronedeploy.com/
17	Enblend	支持正射影像拼接	http://enblend.sf.net
18	Opticks	MSI/HIS 和雷达数据处理	https://opticks.org/
19	Whitebox GAT	水文地貌分析和制图工具	http://www.uoguelph.ca/~hydrogeo/Whitebox/
20	Sentinel Toolbox	Sentinel 3TB 为 Sentinel 任务和第三方数据集提供了大量分析	https://sentinel.esa.int/web/sentinel/toolboxes
21	Diva GIS	生物多样性数据分析、气候数据采集	http://www.diva-gis.org/
22	FalconView	制图和地理参考	https://www.nga.mil/ProductsServices/Pages/-FalconView.aspx
23	Orbis GIS	地理空间数据处理、水文作业	http://orbisgis.org/
24	SAGA GIS:自动化地球科学分析系统	地形处理、分类和摄影测量工具	http://www.saga-gis.org/en/index.html
25	光学和雷达联合地球观测(ORFEO工具箱(OTB))	高空间分辨率数据处理与分析	https://www.orfeo-toolbox.org/
26	地理资源分析支持系统(GRASS)	激光雷达处理和各种空间数据分析	https://grass.osgeo.org/
27	开源软件图像地图(OSSIM)	高性能地理空间数据处理	http://trac.osgeo.org/ossim/
28	综合土地和水信息系统(ILWIS)	RS 数据处理、立体透视和立体浮雕工具	https://52north.org/software/software-projects/ilwis/
29	GvSIG	RS 数据分析	http://www.gvsig.com/en
30	E-foto	摄影测量功能,如立体透视、象片三角测量、三维建模和数字地形校正	http://www.efoto.eng.uerj.br/en

7.6 UAV 开放式多光谱与高光谱数据集研究

空间数据集对于各种应用差异巨大并具有特定性,其中多数都是免费的。卫星图像只能定期获得,其有限的空间分辨率会限制空间目标识别。而 UAV 成像技术已经超越了这一限制;可对任何具有主观测量需求的研究区域进行成像,获得更优的空间数据质量。出于学术目的,可从国家机构和国际机构取得难以采集的数据集或任何特定数据[54]。高分辨率的空间数据大多为商业数据,而政府、军区等敏感区域,只有获得相应的许可才能进行采集[55]。这似乎是开展空间数据研究的一个主要障碍,但相对来说,还有许多资源可以在网上免费提供给公众用户,许多学术研究小组将其成果作为面向公众的开源数据产品[56-58]。大多数 UAV 成像传感器都是商用的,可部署在关注区域进行相应的测量,领先的 UAV 成像公司会将一些数据集作为样本数据集免费提供给用户。可供一般研究团体免费使用的无人机数据资源如下:

- Drone mapper – https://dronemapper.com/sample_data/
- Spectral data sets – http://lesun.weebly.com/hyperspectral – data – set.html
- SPECTIR – https://www.spectir.com/contact
- AusCover airborne data sets – http://www.auscover.org.au/dataset_categories/airborne – datasets/
- CVonline data sets – http://homepages.inf.ed.ac.uk/rbf/CVonline/Imagedbase.htm
- SENSEFLY – https://www.sensefly.com/education/datasets/
- Unmanned aerial RS facility (UARSF) – https://www.wur.nl/en/Research – Results/Chair – groups/Environmental – Sciences/Laboratory – of – Geo – information – Science – and – Remote – Sensing/Research/Integrated – land – monitoring/UARSF.htm
- Rikola hyperspectral data set – http://www.mosaicmill.com/products_other/hs_camera.html

7.7 MSI 和 HIS UAV 成像应用

利用地理空间工具可处理目前获取的各种空间数据,从而得出干旱指数、降水量、气候和悬浮大气微粒方面的数据产品[59-60]。与卫星图像相比,UAV 图像的 MSI 波段映射指数具有更强的鉴别性。这些空间数据衍生产品支持许多应用,具体将在以下各节讨论。

7.7.1 农业监测

采集农田的 UAV 图像可监测作物从发芽到收获这一过程,从而有助于保证各种农业活动的作物健康,甚至可在收获前就通过分析整个生长期的作物健康状况来预测产量,增加收益。乃至土壤肥力、灌溉的水流渠道和肥料利用率也可通过使用 UAV 成像来检查计量。除此之外,由于干旱造成的作物损失也可以通过将损失完全可视化,向保险公司提出合理的索赔[61]。

7.7.2 沿海监视

随着空间数据三维建模技术的不断发展,已进入了通过激光雷达点云来计算水位容量的阶段。机载和星载的空间数据集对于更有效地跟踪水体、河流和其他海岸线的变化极为有利。在特定的需求区域,利用 UAV 图像可很好地规划出海岸体的修复规划措施。这些 UAV 图像也有助于发现非法捕鱼,维护捕鱼法规。在旱天,UAV 图像被用于绘制栖息动物地图,以防止濒危物种灭绝[62]。

7.7.3 林业

无人机在茂密的森林植被方面的应用可有助于政府采取措施,制止非法砍伐和其他侵占行为。例如,阿萨姆邦政府与塔塔咨询服务公司(TCS)合作实施森林防御措施,协助政府采取适当行动,打击猎杀濒危物种(如阿萨姆邦卡齐兰加国家公园的独角犀牛)的偷猎者。利用热成像仪器拍摄的 UAV 图像可通过人员的热特征来定位罪犯。同时,通过应用高效的机器学习技术进行物种识别,可彻查野生动物物种管理。TCS 公司正在阿萨姆邦卡齐兰加国家公园使用深度学习技术进行野生动物物种管理[63]。

7.7.4 城市规划

在城市扩建的情况下,可以利用高分辨率 UAV 图像,通过地面坐标来规划道路、地铁、水道和隧道建设。利用 UAV 成像技术可对矿山和危险场所进行监控,确保工人的安全。在施工现场,为保证进度和实施检查,桥梁或大坝施工等大面积作业,可通过 UAV 图像进行检测。铁路线和水路航道需要大量的人工维护,采用 UAV 成像技术易于测量[64]。

7.7.5 国防应用

目前,对人员居住地至边境线等场所进行监控需要 UAV 来帮助定位可疑活动人员或敌人。在战场上,UAV 可用于侦察偏远或密集地区的敌人。UAV 成像在国防方面的主要优势在于可保护己方不受可疑威胁的袭击,并可用于控诉侵入该国的犯罪者[65]。

7.7.6 环境监测

海岸线、冰川和森林的变化事件都可采用 UAV 图像进行追踪。动植物保护也可采用 UAV 图像进行检查。此外,在自然灾害/各种灾难期间,UAV 图像还可用于灾后评估,以支持政府的救灾措施[66]。

7.7.7 其他商业用途

从考古遗址获得的 UAV 图像既可用于维护历史记载,也可用于跟踪地下新挖掘,进而能节省大量的现场规划和任务执行时间[67]。除 UAV 成像外,这些 UAV 还可能以预定的容量来运送货物,从而有助于保障困难地形战场的后勤。零售巨头亚马孙已迈出了采用无人机投送包裹的步伐,并向相关创企提供了资金支持[68]。

7.8 结论和未来展望

利用 UAV 进行环境监测当前极具吸引力,因此提高稳定性和承压能力极为重要,从而可通过携带高度专业化的成像仪来设计有效的开发方法。与现有的 MSI 仪器相比,HSI 技术在多个通道中详细描述了光谱信息,以更高的分辨率分析了潜在的空间现象。MSI 是遥感技术中最有前景的技术,其仪器性能更接近于地面传感,但大数据无序采集处理和瞬时图像处理仍是公开的挑战。

MSI 和 HSI 传感器已使 RS 成为在农业、医药、环境保护、矿物学、减灾、物理和监视等诸多领域的重要决策工具之一。然而,获取的高非线性和高维图像为潜在用户带来了新的警示、挑战和不可预见的限制。近年来,软计算技术因其具有加速优化和处理非线性数据的能力而成为一种可选方案,并引起了不同知识领域的专家、管理人员和团体成员的广泛关注,从而为解决其中部分问题提供了可能。

对这些资源进行充分解读是非常必要的。通常,在不同空间分辨率的多幅传感器卫星图像中检测特定区域或单个目标是 RS 成像分析中的一个关键问题。使用传统的分类器对这类目标检测需要大量的训练时间和繁琐的 GT 标签。因此,需要新的范式来减少复杂性,并采用如加速稳健特征(一种广泛使用的稳健局部特征描述符)所形成的描述符来对所选地面覆盖区域进行高性能检测。特征向量的维数很大程度上取决于数据的表示、运行和存储方式[20,69]。鉴于GT概念也与维数问题密切相关,图像描述符会产生图像语义处理问题,这可能是工业5.0带来的最大的变化之一。

计算智能可直接或协助解决专家们所遇到的一些试验。同时,它们推动了RS和计算机视觉在CPS中的应用创新以扩展加快,如变化检测、全色锐化、波段选择、分割、图像分解、聚类、分类、端元提取和语义鸿沟缩小等任务[49,69-72]。

语义网又称数据网,旨在实现机器的互操作性。目前,Web服务基于Web服务定义语言(WSDL),它是一种XML格式,将服务定义为处理消息的一组端点,包括面向文档或面向过程的信息。通常,服务是接受Web协议的一组系统功能的容器。类似地,在Web服务中,客户机读取WSDL文件来查找服务器提供的操作。然后简单对象访问协议帮助调用WSDL中列出的其中一个操作。在语义映射上,可将这些WSDL服务映射至OWL-S服务,这可以通过添加语义注释来完成,人工操作是一件令人厌烦的工作。采用OWL-S服务可实现语义映射自动化,过去采用DARPA代理标记语言的语义映射方法是半自动的,许多基于OWL-S服务的语义映射系统都是全自动化的[18-19]。

将UAV系统建模为CPS既可应对新的需求,也可满足物联网、软件定义无线电和大数据等蓬勃发展的领域,因为如果不利用这些领域的优势,发展分布式多无人机是不可能取得成功的[70,73-75]。

参考文献

[1] Google Self Driving Car-[Online]. https://www.thetimes.co.uk/article/google-waymo-car-of-the-future-is-driving-residents-mad-vnkwpjg7q [Accessed 21 Sep 2018].

[2] Tesla Self Driving Feature-[Online]. https://www.wired.com/story/tesla-self-driving-car-computer-chip-nvidia/ [Accessed 21 Sep 2018].

[3] T. Ishida, J. Kurihara, F. A. Viray, et al. 'A novel approach for vegetation classification using UAV-based hyperspectral imaging'. *Computers and Electronics in Agriculture*. 2018; 144: 80-85.

[4] Yong-Gu Han, Se-Hoon Jung, and Ohseok Kwon. 'How to utilize vegetation survey using drone image and image analysis software'. *Journal of Ecology and Environment*. 2017; 41: 1-6.

[5] M. A. Khan, W. Ectors, T. Bellemans, D. Janssens, and G. Wets. 'UAV-based traffic analysis: a universal guiding framework based on literature survey'. *Transportation Research Procedia*. 2017; 22: 541-550.

[6] Electromagnetic Spectrum- https://nl.wikipedia.org/wiki/Bestand:EM_Spectrum3-new.jpg [Accessed 22 Sep 2018].

[7] K. Uto, H. Seki, G. Saito, and Y. Kosugi. Development of Lightweight Hyperspectral Imaging System for UAV Observation. 2014 6th Workshop on Hyperspectral Image and Signal Processing:

Evolution in Remote Sensing. 2014;1-4.

[8] H. Kwon, X. Hu, J, Theiler, A. Zare, and Prudhvi Gurram. 'Algorithms for multispectral and hyperspectral image analysis, *Journal of Electrical and Computer Engineering*. Vol 2013, 2013, pp. 908906:1-908906:2.

[9] A. Lucieer, Zbyněk Malenovský, Tony Veness, and Luke Wallace. 'HyperUAS – imaging spectroscopy from a multirotor unmanned aircraft system'. *Journal of Field Robotics*. 2014; 31(4): 571-590.

[10] R. Mohan, C. Vipin Raj, P. Aswathi, and Rao R. Bhavani. 'UAV based security system for prevention of harassment against woman'. *Proc. Intelligent Computing Instrumentation and Control Technologies (ICICICT)*. Kannur, India, 2017; 874-879.

[11] Sicong Liu, Qian Du, Xiaohua Tong, Alim Samat, Haiyan Pan, and Xiaolong Ma. 'Band selection – based dimensionality reduction for change detection in multi – temporal hyperspectral images'. *Remote Sensing*. 9(10), 1008, 2017.

[12] H. Huang, F. Luo, J. Liu, and Y. Yang. 'Dimensionality reduction of hyperspectral images based on sparse discriminant manifold embedding'. *ISPRS Journal of Photogrammetry*. 2016; 106: 42-54.

[13] J. Meszaros, Aerial Surveying Uav Based On Open – Source Hardware And Software. Proc. International Archives of the Photogrammetry, Remote Sensing and Spatial Information Sciences, Volume XXXVIII – 1/C22, 2011 ISPRS Zurich 2011 Workshop, 14 – 16 September 2011, Zurich, Switzerland.

[14] UAV Sense of Direction – [Online]: http://www.asctec.de/en/ethz-drones-with-a-sense-of-direction/ [Accessed 18 Sep 2018].

[15] UAV Design – [Online]: https://en.wikipedia.org/wiki/Unmanned_aerial_vehicle [Accessed 28 Aug 2018].

[16] S. Huh, D. H. Shim, and J. Kim. 'Integrated Navigation System Using Camera and Gimbaled Laser Scanner for Indoor and Outdoor Autonomous Flight of UAVs'. 2013 IEEE/RSJ International Conference on Intelligent Robots and Systems. Tokyo, Japan, 2013, 3158-3163.

[17] Drone – [Online]: http://www.i-runway.com/blog/drones-hitch-rides-scream-relax/ [Accessed 20 Sep 2018].

[18] R. J. Aroma, and M. Mathew Kurian. A Semantic Web: Intelligence in Information Retrieval. *Proc. 2013 IEEE International Conference on Emerging Trends in Computing, Communication and Nanotechnology (ICECCN 2013)*. Tirunelveli, India, 2013.

[19] R. J. Aroma, and M. Mathew Kurian. 'A survey on tools essential for semantic web research'. *International Journal of Computer Applications (ISSN 0975-8887)*. 2013;62(9):26-29.

[20] R. J. Aroma, and K. Raimond. (2019). Intelligent Land Cover Detection in Multi – sensor Satellite Images. In: Ane B., Cakravastia A., and Diawati L. (eds) Proceedings of the 18th Online World Conference on Soft Computing in Industrial Applications (WSC18). WSC 2014. Advances in Intelligent Systems and Computing, World Wide Web, vol. 864: 118-128. Springer, Cham, Zurich, Switzerland.

[21] Shefali Aggarwal. Principles of Remote Sensing, Satellite Remote Sensing and GIS Applications in Agricultural Meteorology. *Proceedings of the Training Workshop*. Dehra Dun, India, 2003, pp. 39 – 65.

[22] Resolution of Remote Sensing – [Online]: www. edc. uri. edu/nrs/classes/NRS409/RS/Lectures/HowRemoteSensonWork. pdf [Accessed 21 Sep 2018].

[23] A. M. Melesse, Q. Weng, P. S. Thenkabail, and G. B. Senay. 'Remote sensing sensors and application in environmental resource mapping and modelling'. Sensors. 2007; 7(12).

[24] Impact of Spatial Resolution in Digital Images – [Online]: http://micro. magnet. fsu. edu/primer/java/digitalimaging/processing/spatialresolution/ – [Accessed 20 Sep 2018].

[25] J. R. B. Fisher, E. A. Acosta, P. James Dennedy – Frank, T. Kroeger, and T. M. Boucher, 'Impact of satellite imagery spatial resolution on land use classi – fication accuracy and modeled water quality'. *Remote Sensing in Ecology and Conservation*. 2018; 4(2):137 – 149.

[26] S. A. Boyle, C. M. Kennedy, J. Torres, K. Colman, P. E. Perez – Estigarribia, and Noe U. de la Sanch. 'High – resolution satellite imagery is an important yet underutilized resource in conservation biology'. *PLoS One*. 2014 Jan 23; 9(1): e86908. DOI: 10. 1371/journal. pone. 0086908.

[27] B. Krishna Mohan, and A. Porwal. 'Hyperspectral image processing and analysis'. *Current Science*. 108(5), 833 – 841.

[28] Hyperspectral Remote Sensing – [Online]: http://www. csr. utexas. edu/project/rs/hrs/hyper. html [Accessed 20 Sep 2018].

[29] Multispectral vs. Hyperspectral Imagery Explained – [Online]: http://gis – geography. com/multispectral – vs – hyperspectral – imagery – explained/ [Accessed 20 Sep 2018].

[30] I. Vorovencii. 'Optical and active satellite sensors used in forestry'. *Bulletin of Transilvania University of Brasov*. 2012; 5: 63 – 69.

[31] Hyperspectral Imagery – [Online]: https://www. cerc. usgs. gov/morap/Assets/UploadedFiles/Projects/Contamination_Characterization_through_Airborne_Hyperspectral_Imaging/What_Is_Hyperspectral_Imaging. pdf [Accessed 21 Sep 2018].

[32] Michael T. Eismann, Hyperspectral Remote Sensing. SPIE, Bellingham, Washington, USA. DOI: 10. 1117/3. 899758. PDF ISBN: 9780819487889 | Print ISBN: 9780819487872

[33] E. Underwood, S. Ustin, and D. DiPietro. 'Mapping non – native plants using hyperspectral imagery'. *Remote Sensing of Environment*. 2003; 86(2):150 – 161.

[34] T. R. Martha, K. BabuGovindharaj, and K. Vinod Kumar. 'Damage and geological assessment of the 18S eptember 2011 Earthquake in Sikkim, Indiausing very high resolution satellite data'. *Geoscience Frontiers*, 6, 6, 2014, pp. 793 – 805.

[35] Satellite Images – Online: http://en. wikipedia. org/wiki/Satellite_imagery [Accessed 21 Sep 2018].

[36] X. Xiaohong, and W. Yonggang. 'A cloud – removal method based on image fusion using local indexes'. *Computer Modelling & New Technologies*. 2014; 18(4): 82 – 88.

[37] A. Rassoul, S. Mahiny, J. Brian, and A. Turner. 'Comparison of four com – mon atmospheric correction methods'. *Photogrammetric Engineering & Remote Sensing*. 2007; 4:361 – 368.

[38] O. I. Singh, T. Sinam, O. James, and T. R. Singh. 'Local contrast and Mean based thresholding technique in image binarization.' *International Journal of Computer Applications.* 2012;51(6): 6–10.

[39] A. Makandar, and B. Hallali. 'Image enhancement techniques using high–pass and low pass filters.' *International Journal of Computer Applications.* 2015; 109(14):12–15.

[40] Gamma Correction–[Online]: http://xahlee.info/img/what_is_gamma_cor–rection.html [Accessed 23 Sep 2018].

[41] Spectral Reflectance Graph–[Online]: http://www.seos–project.eu/modules/classification/classification–c01–p05.html [Accessed 21 Sep 2018].

[42] Red shift–[Online]: https://en.wikipedia.org/wiki/Redshift [Accessed 21 Sep 2018].

[43] M. Ghazal, Y. Al Khalil, and H. Hajjdiab. UAV–based Remote Sensing for Vegetation Cover Estimation Using NDVI Imagery and Level Sets Method. *IEEE International Symposium on Signal Processing and Information Technology.* 2015; 332–337.

[44] T. Sakamotoa, A. A. Gitelsonb, A. L. Nguy–Robertsonb, T. J. Arkebauer, B. D. Wardlow, A. E. Suyker, S. B. Verma, M. Shibayama. 'An alternative method using digital cameras for continuous monitoring of crop status.' *Agricultural and Forest Meteorology.* 2012; 154–155: 113–126.

[45] E. R. Hunt, C. S. T. Daughtry, S. B. Mirsky, and W. D. Hively. 'Remote sensing with simulated unmanned aircraft imagery for precision agriculture appli–cations.' *IEEE Journal of Selected Topics in Applied Earth Observations and Remote Sensing.* 2014; 7: 4566–4571.

[46] S. Mondal, C. Jeganathan, N. K. Sinha, H. Rajan, T. Roy, and P. Kumar. 'Extracting seasonal cropping patterns using multi–temporal vegetation indices from IRS–LISS III Data in Muzaffarpur District of Bihar, India'. *The Egyptian Journal of Remote Sensing and Space Science.* 2014;17(2),123–134.

[47] J. Valasek, J. V. Henrickson, E. Bowden, Y. Shi, C. L. S. Morgan, and H. L. Neely. 'Multispectral and DSLR sensors for assessing crop stress in corn and cotton using fixed–wing unmanned air systems'. *Proceedings Volume 9866, Autonomous Air and Ground Sensing Systems for Agricultural Optimization and Phenotyping*; 98660L, SPIE Commercial + Scientific Sensing and Imaging, Baltimore, Maryland, USA, 2016. doi: 10.1117/12.2228894

[48] Jinya Su, Dewei Yi, Cunjia Liu, Lei Guo, and Wen–Hua Chen. 'Dimension reduction aided hyperspectral image classification with a small–sized train–ing dataset: experimental comparisons'. *Sensors.* 17(12),2726,2017

[49] A. M. Coelho, and V. V. Estrela. EM–based Mixture Models Applied to Video Event Detection. In: Sanguansat P (ed) Principal Component Analysis—Engineering Applications. InTech Open, London, UK, 2012, doi: 10.5772/38129.

[50] F. D. Silva, V. V. Estrela, and L. J. Matos. (2011). Hyperspectral Analysis of Remotely Sensed Images, In: C. Bilibio, O. Hensel, and J. F. Selbach (Eds), Sustainable Water Management in the Tropics and Subtropics and Case Studies in Brazil, vol. 2, ISBN 9788563337214, pp. 398–423, Kassel, Germany.

[51] Y. Bazi, and F. Melgani. 'Convolutional SVM networks for object detection in UAV imagery.'

IEEE Transactions on Geoscience and Remote Sensing, 56, 6, pp. 3107 – 3118, 2018.

[52] Dinale Zhou, Jinglun Zhou, Maojun zhang, Dao xiang, and Zhiwei Zhong. 'Deep Learning for Unmanned Aerial Vehicles Landing Carrier in Different Conditions', Proc. 2017 18th Int'l Conf. on Advanced Robotics (ICAR), Hong Kong, China, 2017, 469 – 475.

[53] U. Niethammer, S. Rothmund, U. Schwaderer, J. Zeman, and M. Joswig. Open Source Image – Processing Tools for Low – Cost UAV – Based Landslide Investigations. International Archives of the Photogrammetry, Remote Sensing and Spatial Information Sciences, Volume XXXVIII – 1/C22, 2011. ISPRS Zurich 2011 Workshop, 14 – 16 September 2011, Zurich, Switzerland, 2011, 14 – 16

[54] NRSC Data Product Services and Regulations – [Online]: http://www.nrsc.gov.in/Data_Products_Services_Satellite_Data_Order.html [Accessed 21 Sep 2018].

[55] Quick Bird – Commercial data – [Online]: https://www.satimagingcorp.com/satellite – sensors/quickbird/ [Accessed 21 Sep 2018].

[56] Statlog (Landsat dataset) in UCI Machine Learning Repository – Online: https://archive.ics.uci.edu/ml/datasets/Statlogt(LandsattSatellite) [Accessed 21 Sep 2018].

[57] Arizona State University (ASU) Mars Global Data – Online: http://www.mars.asu.edu/data/ [Accessed 21 Sep 2018].

[58] Space Science and Engineering Center at University of Wisconsin – Online: http://www.ssec.wisc.edu/data/ [Accessed 21 Sep 2018].

[59] S. M. Kotikot, and S. M. Onywere. 'Application of GIS and Remote sensing techniques in frost risk mapping for mitigating agricultural losses in the aberdare ecosystem, kenya'. Geocarto International. 2015; 30.

[60] B. Gharai, S. Jose, and D. V. Mahalakshmi. 'Monitoring intense dust storms over the Indian region using satellite data – a case study'. International Journal of Remote Sensing. 2013; 34: 7038 – 7048.

[61] M. Reinecke, and T. Prinsloo. 'The influence of drone monitoring on crop health and harvest size'. Proc. 1st International Conference on Next Generation Computing Applications (NextComp), pp. 5 – 10. IEEE, 2017

[62] J. A. Goncalves, and R. Henriques. 'UAV photogrammetry for topographic monitoring of coastal areas'. ISPRS Journal of Photogrammetry and Remote Sensing. 2015; 104: 101 – 111. DOI: 10.1016/j.isprsjprs.2015.02.009

[63] T. P. Banu, G. F. Borlea, C. Banu. 'The use of drones in forestry.' Journal of Environmental Science and Engineering. 2016; 557 – 562.

[64] Lei Zhu, Dong Yin, Lincheng Shen, Xiaojia Xiang, and Guanghua Bai. Research on Urban Application – oriented Route Planning of UAV Based on Mobile Communication Network. Proc. International Conference on Computer Science and Network Technology (ICCSNT 2015). Vol. 1, 2015; 1562 – 1570.

[65] A. Sehrawat, T. Choudhury, and G. Raj. Surveillance Drone for Disaster Management and Military Security. 2017; 470 – 475.

[66] G. F. Griffin. 'The use of unmanned aerial vehicles for disaster manage‐ment'. *Geomatica*. 2014; 68(4): 265-281.

[67] Michael Doneus, Geert Verhoeven, Clement Atzberger, Michael Wess, and Michal Ru. 'New ways to extract archaeological information from hyper‐spectral pixels'. *Journal of Archaeological Science*. 2014; 52: 84-96.

[68] Amazon Drone Delivery ‐ [Online]: https://en.wikipedia.org/wiki/Amazon_Prime_Air [Accessed 21 Sep 2018].

[69] A. E. Herrmann, and V. V. Estrela. (2016). Content‐Based Image Retrieval (CBIR) in Remote Clinical Diagnosis and Healthcare. In: Cruz‐Cunha M., Miranda I., Martinho R., and Rijo R. (eds) Encyclopedia of E‐Health and Telemedicine. pp. 495-520, IGI Global. Hershey, PA, USA. doi: 10.4018/978-1-4666-9978-6.ch03.

[70] D. J. Hemanth, and V. V. Estrela. Deep Learning for Image Processing Applications, Advances in Parallel Computing Series, vol. 31, IOS Press, Netherlands, ISBN 978-1-61499-821-1 (print), ISBN 978-1-61499-822-8 (online).

[71] M. A. de Jesus, V. V. Estrela, O. Saotome, and D. Stutz. (2018). Super‐Resolution via Particle Swarm Optimization Variants. In: Hemanth J., and Balas V. (eds) Biologically Rationalized Computing Techniques for Image Processing Applications. LNCVB, vol. 25. Springer, Cham, Zurich, Switzerland.

[72] S. R. Fernandes, V. V. Estrela, H. A. Magalhaes, and O. Saotome. (2014). On Improving Sub‐pixel Accuracy by Means of B‐Spline. *Proceedings of the 2014 IEEE International Conference on Imaging System and Techniques (IST 2014)*. Santorini, Greece. doi: 10.1109/IST.2014.6958448.

[73] V. V. Estrela, O. Saotome, H. J. Loschi, D. J. Hemanth, W. S. Farfan, R. J. Aroma, C. Saravanan, and E. G. H. Grata, 'Emergency response cyber‐physi‐cal framework for landslide avoidance with sustainable electronics'. Technologies. 2018; 6: 42. doi: 10.3390/technologies6020042.

[74] V. V. Estrela, J. Hemanth, O. Saotome, E. G. H. Grata, and D. R. F. Izario, Emergency response cyber‐physical system for flood prevention with sus‐tainable electronics. In: Iano Y., Arthur R., Saotome O., Estrela V. V., and Loschi H. (Eds) Proceedings of the 3rd Brazilian Technology Symposium. BTSym 2017. Campinas, SP, Brazil, Springer, Cham, Zurich, Switzerland, 2019, pp. 319-328. DOI: 10.1007/978-3-319-93112-8_33

[75] V. V. Estrela, A. C. B. Monteiro, R. P. Franc. a, Y. Iano, A. Khelassi, N. Razmjooy, Health 4.0: Applications, Management, Technologies and Review. Medical Technologies Journal, 2019; 2 (4):262-76, DOI: 10.26415/2572-004X-vol2iss1p262-276.262.

[76] V. V. Estrela, A. Khelassi, A. C. B. Monteiro, Y. Iano, N. Razmjooy, D. Martins, and D. T. M. Rocha. "Why software‐defined radio (SDR) matters in healthcare?". Medical Technologies Journal, vol. 3, no. 3, Oct. 2019, pp. 421-429, doi:10.26415/2572-004X-vol3iss3p421-429.

[77] H. Mei, K. Wang, D. Zhou, and K. Yang, Joint Trajectory‐Task‐Cache Optimization in UAV‐Enabled Mobile Edge Networks for Cyber‐Physical System. IEEE Access, 7, 156476-156488, 2019.

第8章 无人机航空成像与基础设施重构

本章介绍利用微型飞行器(MAV)进行大规模基于视觉的基础设施重构的试验性实地试验的汇编,研究的重点是传感器的选择、数据集的生成以及三维模型生成的计算机视觉算法。一般来说,MAV以其在各种速度下飞行、稳定位置以及在大型基础设施附近执行机动的能力而著称,上述优点使空中机器人成为一个用于基础设施检查和维护任务的高速发展的机器人平台,可以开发具有面向任务的感官模式的不同MAV解决方案来执行独特的任务,如基础设施的三维建模。在本章中,空中机器人在不同的环境中导航,同时使用运动恢复结构(SFM)和多视图立体(MVS)技术收集视觉数据并进行后置处理,以生成三维模型[1-2],该框架在室内、室外和地下环境中已成功进行了试验验证。

8.1 引言

微型空中机器人在受限且定义明确的试验室环境中表现出更强的稳健性和技术性能。然而,这项技术已经进入一个新的时代,设想在真实规模的基础设施环境中部署,并具有展示高度自治水平的能力[3-4]。

与地面移动机器人相比,MAV的独特之处在于其导航模式的多样性,从在目标上空盘旋到在笛卡儿坐标系中进行攻击性部署,这些能力使它们能够进入偏远和难以到达的地区。此外,MAV的独特之处在于它们的有效载荷能力和它们的尺寸之间的权衡。通常情况下,它们配备了相机、声纳和激光等传感器以及GPS,为他们的操控员提供有用的信息。因此,这些平台的卓越能力不断达到新的技术壁垒,并缓慢而稳定地确立了空中机器人在具有巨大社会和经济扩展的工业运营中的地位[5]。

此外,在具有挑战性或恶劣环境中执行任务时,为MAV配备各种传感器套件对于广泛的应用是有帮助的,这些功能适用于复杂的任务执行,这些任务不可能由人工执行,也不需要人工执行,因此减少了总体执行时间,提高了检查任务的效率(尤其是与传统方法相比)[6]。到目前为止,它们已经被集成到手动驾驶的摄影拍摄行业中,但是,越来越多的资源被投入到远程自主检查应用中。使用MAV的最新成果包括基础设施检查[7-8],公共安全,如监视[9]和搜索救援任务[10]。总的来说,最近,采矿、石油和能源供应商等行业都在大力投资MAV的商业化,以执行远程检查应用[11-12]。这些努力将产生即将来临的影响,改变目前在基础设施检查领域使用的普遍做法。例如:① 整体基础设施运营;② 生产;③ 安全。

在这些操作场景中,MAV被分配了半自主导航任务,以进行数据采集。数据随车载传感器的不同而变化,可以是可视的、热传感的、距离的、气体的以及其他类型的传感器。在这项研究中,重点是分析视觉数据,用于基础设施/检查区域的三维表面重构。鉴于此,在所提出的现场试验中,主要应用场景是在目标区域周围的单个或多个空中平台的自主导航、可视化数据采集,同时应用离线后置处理方法来创建三维模型。需要创建被检查表面的三维模型源于需要进行体积分析、损伤缺陷分析,最重要的是在已知缺陷位置时创建参考模型,并且空中平台可以在另一时间重复检查任务。此外,当任务正在进行时,资产分析师可以直接使用可视流对访问的区域进行初步检查。

任务要求、结构物的几何形状和结构物的外观是影响三维模型是在线生成还是离线生成的主要因素。通常,深度传感相机,包括立体传感器、RGB-D传感器和单目相机,在检查任务中提供足够信息的同时,使用不同的视觉算法处理方法。当结构物是大尺度时,一个重要的问题是来自RGB-D相机或立体相机的深度传感信息会被大幅限制,使得这些传感器的性能类似于单目相机。鉴于此,立体算法无法在结构物与相机基线相比处于更大距离的情况下执行[13],或者在无纹理的环境中执行。

8.2 相关研究

尽管空中机器人的研究在定位[14]、规划[15]和感知[16]方面已经达到了重要的里程碑,但在文献中很少有关于自动检查系统的成功的实际演示报道。大多数应用集中于在完全控制环境下进行的令人印象深刻的试验室试验,而且在大多数情况下,是在使用昂贵的运动捕捉系统[17]或小规模和明确的室外环境[18-19]下进行的。在参考文献[20]中,提出了一种在风力发电机模拟环境中进行检查的空中系统。这项研究展示了一个原型平台,该平台配备激光雷达和GPS,具有三维测绘能力,并配备了无人机的激光雷达传感器和防撞距离控制系统。在参考文献[21]中,提出了一个空中平台,用于收集未铺砌道路状况的数据集。在参考文献[22]中,进行了一项比较研究,以测试在沿海环境中从低空空中平台收集的数据集中使用 SfM 的情况。

在地下矿山作业中的 MAV 的相关文献中,很少有研究试图解决这些具有挑战性的任务。在参考文献[23]中,提出了一种视觉惯性(VI)导航框架来实现平台的位置跟踪控制,在这种情况下,MAV 被控制飞行在无障碍路径上,而系统在真实的隧道环境中进行了试验评估,模拟了一个煤矿矿场,在那里假设照明问题已经解决。在参考文献[24]中,与参考文献[23]中有关地下定位的方法相比,采用了更为现实的方法,更具体地说,一架装有 VI 传感器和激光扫描仪的六轴飞行器被人工引导穿过垂直矿井,以收集数据进行后置处理,从测量中提取的信息以用于创建环境的三维网格并定位车辆。最后,在参考文献[25]中,使用惯性测量单元(IMUS)、相机和激光雷达传感器,研究了使用飞行器自主检查压力管道和隧道的估计、导航、测绘和控制能力。此外,针对无人地面车在地下矿山的导航和勘探,参考文献[26]提出了一种地面机器人用于隧道网络中的自主勘探。他们的系统有两种操作模式:第一种模式下,操控员远程移动平台并在绘制地图的同时收集距离数据;第二种模式下,上级任务规划人员提供适当的导航操作。文献作者假设环境的连通图是已知的,并且操控员在该图上定义了路径。此外,假设了系统的完整里程计和地图是已知的。在参考文献[27]中,开发了一种带有机械手的地面机器人,用于在已知的地下环境地图中进行自动导航。该平台配备了 RGB-D 相机、光源、激光扫描仪和 IMU,并在考虑环路闭合的情况下绘制了该区域的地图。在参考文献[28]中,研究了一个地面机器人团队在受限和结构化环境中进行探测的情况。多机器人自主运动规划和导航、定位、映射以及机器人之间的实时通信是所开发框架的功能,在只有特征或自然地标被认为是已知的情况下[29],用于执行无人地面车的手动操作。此项研究在矿山和铁路隧道环境中进行了试验评估。在参考文献[30]中,一辆汽车装备了一个二维激光扫描仪、工业级 IMU,并且传感器以一个六自由度的轨迹通过环境。在人工驾驶车辆的同时,提取了 17km 的矿井隧道地图。

Wang 等[31]开发并设计了一种移动机器人来代替煤矿救援任务中的人。该平台具有防爆、防水的特点,同时对通信技术进行了研究,并在一个煤矿矿场对该平台进行了评估。虽然基于地面无人车或汽车的研究表明其在地下环境中具有良好的性能,但这些方法不能直接应用于 MAV,而且大多数是人工操作的。由于 MAV 的有效载荷能力有限,不能携带所有地面无人车传感器套装,并且在位置估计时会发生漂移,而且这些平台的快速动态特性,会导致碰撞。此外,地面机器人在泥浆、水和坑洼中的导航和定位是极其复杂的任务[32]。然而,这些因素可能不会影响 MAV,因为它完美避开了这些导致性能下降的因素。

8.3 视觉传感器和任务规划器

如今,嵌入式系统和微型化的发展带来了功能强大且成本低廉的相机模块,这些模块可以安装在 MAV 上,并结合经典导航设备的数据来反馈必要的视觉信息。在这一节中,简要讨论了相机投影模型,重点介绍了相机的操作,同时介绍了基于相机投影模型的路径规划器的设计。

8.3.1 图像投影

构造图像框架:首先将三维信息投影到二维空间,去除真实世界中的深度信息,这一过程称为透视投影;然后,利用针孔相机投影模型将点从二维空间映射到具有像素坐标的点二维图像平面,形成数字图像。

在计算机视觉任务中,通常采用中心透视成像模型(图 8.1)。光线会聚在相机坐标系 C 的原点上,并且非反转图像被投影到位于 $z=f$ 的图像平面上,其中 f 称为焦距。利用相似三角形原理,一个在世界坐标系的点 $P=(X,Y,Z)$ 被投影到图像平面 $p=(x,y)$,其中 $x=f\dfrac{X}{Z}$ 和 $y=f\dfrac{X}{Z}$。此外,在视觉传感器中,通常还要进行一个步骤,即投影到数字图像空间。由于一些原因,这种投影并不直接对应于数字图像中的像素坐标。通常,图像平面中的像素坐标系不同于由平移向量 $[c_x,c_y]$ 偏移的参考系。另外,数字图像被划分为离散像素,而相机坐标系中的点是连续的,用像素坐标中的 f_x 和 f_y 表示。物理传感器会引入非线性因素,如映射失真。通过引入 $P_{\text{digital}}=(u,v)$,其中 $u=f_x XZ+c_x$,$v=f_y YZ+c_y$,对投影模型引入额外的变换来考虑所有上述差异。

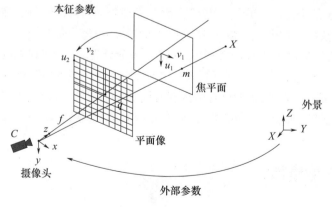

图 8.1 图像投影示意图

8.3.2 路径规划器

为了正确利用视觉传感器的投影模型并能够重构三维表面,所开发的框架应结合导航功能,这些导航功能应考虑相机规格,并在连续捕捉图像中收集具有所需视场(FOV)的视觉数据。在本章中,MAV 使用[7]中建立的覆盖方案在结构物周围导航。

对于路径规划器的建立,最初考虑了配备有限 FOV 传感器的机器人的一般情况,由孔径角 α 和最大范围 r_{max} 确定,如图 8.2 所示。此外,$\Omega \in R^+$ 是用户定义的距基础结构目标曲面的偏移距离($\Omega < r_{max}$),$\Delta \lambda$ 是每个切片平面之间的距离,$\Delta \lambda = \dfrac{\Omega}{2} \tan \dfrac{x}{2}$,如图 8.2 所示。

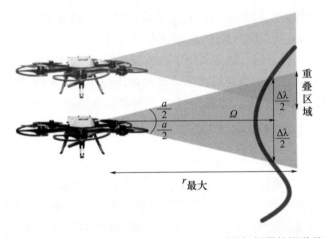

图 8.2 考虑配备有限 FOV 传感器的机器人覆盖规划器的视觉足迹

此外,该路径规划器为 MAV 提供覆盖基础设施的轨迹,将结构物离散化为水平面,这是分支标识的一部分,并最终为每个代理分配特定的区域。在以不同分支为特征的复杂结构物中,结构物被切分成更小的部分,每个部分视为一个分支。例如,对于风力发电机,基座和每一个叶片视为一个分支,其中规划人员提取这些分支并将轨迹分配给 n 个代理。规划人员考虑到视觉传感器的特性,并使用此信息来确保完全覆盖,因为轨迹为所有结构部件提供重叠视图。最后,规划器的另一个优点是,它提供偏航参考和位置参考,以指导 MAV 始终朝向被检查表面。

8.4 三维重构

本章的一部分旨在描述用于生成三维模型的方法,同时处理在任务期间收集的数据。最初,考虑了两种主要方法,即使用立体或单目相机映射,这两种方法之间的选择主要取决于应用的需求和对象的结构/外观。

在本研究中,这两种方法都用作概念证明,并将在8.6节中提出。

8.4.1 立体映射

对于立体三维映射场景,每个机器人在其局部坐标框架中提供一个姿态估计和一个被检查表面的三维模型。在已开发的情况下,代理将自己定位在全局定位系统(例如,运动捕捉系统)的坐标框架中,并将该里程计信息集成到立体SLAM[33]算法中。从各个代理生成的所有点云都在坐标框架中表示。当部署 n 个代理时,每个代理覆盖目标对象周围的特定区域,重构特定部分。此后,通过迭代最近点(ICP)[34]算法合并每一个由 n 个代理创建的映射,以形成整个结构的全局表示。该技术以两个点集 E 和 Z(算法1)作为输入,使用一组经过滤波的对应点 $f(E)$ 和 $f(Z)$ 来计算表示它们之间相对姿态的三维变换 τ。首先通过最小化误差 $E_{er} = \sum_{j=1}^{N} (\| \tau(\xi_j) - \xi_i \|^2)$ 来估计变换;然后应用 τ 来校正公共帧中的点集。输出是一个全局点集 M,表示整个场景的三维重构。因此,代理覆盖场景的公共部分以校正各个绘制地图是很重要的,而该过程是离线执行的。

为了生成三维模型,采用了最先进的映射算法,称为 RTABMap[35] SLAM,该算法适用于大规模操作,同时基本上解决了复杂结构的映射。更具体地说,RT-ABMap是一种基于外观的定位和映射算法,它由三部分组成:视觉里程计、回环检测和图形优化部分。

首先,该算法识别图像中的特征,并为环路闭合处理建立一个可视化词汇表[36]。此过程用于确定以前是否访问过新的图像位置。然后,使用深度测量值将特征位置从图像投影到三维,并通过过滤步骤去除异常值。如果特征在前一帧中被识别,则该帧将作为节点添加到姿态图中,而贝叶斯滤波器会跟踪回环情况。当回环被识别时,图像位置被添加到保存映射 M_i 的图中。最后,进行位姿图优化和光束平差以细化估计的位置。上述过程在算法1中进行了描述。

算法1　Stereo 3D reconstruction

1: function ICP FRAME ALIGNMENT
2: Map 1 point set $E = \xi_1, \xi_2, \xi_3, \cdots, \xi_N$
3: Map 2 point set $Z = \xi_1, \xi_2, \xi_3, \cdots, \xi_N$
4: Extract point features(Ξ)→$f(\Xi)$
5: Extract point features(Z)→$f(Z)$
6: Match $f(\Xi)$ and $f(Z)$

7：Filter false matches and remove outliers
8：$\min E_{er} = \sum_{j=1}^{n}(\|\tau(\xi_j) - \xi_i\|^2)\tau$
9：Apply transform τ and align point sets
10：end function
11：function VISUAL SLAM
12：Frame location initialisation (Feature Detection, Visual word vocabulary)
13：Loop closure detection
14：Frame location to pose graph
15：Pose graph optimization
16：Point cloud with corresponding pose→M
17：end function Assume to have n agents
18：for agent i do
19：Visual SLAM→M_i；3D map generated for each agent
20：end for
21：Merge Maps
22：for i：1：#$maps$ do
23：ICP FRAME ALIGNMENT(M_i)→global map M；Agents with common scene coverage
24：end for

8.4.2 单目映射

在单目映射的情况下，增量 SFM[1,2,37,38] 技术用于创建被检查对象的三维模型。在执行任务期间，空中代理沿着目标表面导航，按照指定的路径将单目相机的图像流存储在机载存储器中。

将采集到的数据集存储在离线使用不同相机视点的数据库中，并通过 SFM 算法进行处理。首先，该算法执行对应搜索步骤，识别输入图像间重叠的场景部分，在此阶段，进行特征提取和帧间匹配算法，提取图像场景覆盖信息；然后使用对极几何[39]进行几何验证以消除错误匹配。在这种方法中，在增量注册新帧之前，选择具有足够视差的初始图像对 I_1 和 I_2 以执行双视图重构是至关重要的。

SFM 算法：首先恢复两幅图像中的匹配特征集 f_1 和 f_2；然后使用五点算法[40]估计 I_1 和 I_2 的相机外参，用奇异值分解（SVD）分解结果基本矩阵 E_{es}；其次构建包含每个帧的旋转估计和平移估计的投影矩阵 $P_i = [R_i | t_i]$；最后，利用相对姿态信息，对识别出的特征进行三角剖分，恢复其三维位置 X^{3D}。两帧光束平差负责细化已识别特征在图像帧中的位置，使重投影误差最小化。在此初始化步骤之后，剩余的图像将依次注册到当前相机和点集中。更具体地说，捕捉到最多恢复三维点的帧由透视 n 点（PnP）[41]处理。该算法利用二维特征点与三维特征

点的对应关系来提取它们的姿态。此外,新注册的图像将使用多视图三角剖分扩展现有的三维场景集(X^{3D})。下面对整个模型进行全局光束平差,以校正过程中的漂移。上述过程在算法 2 中进行了描述。在覆盖任务期间,代理基于 VI 里程计自主飞行(8.6 节)。通过在过程中结合附加信息,使用来自相机机载定位的全姿态注释图像或来自环境的任何其他已知度量信息,如校准模式,可以恢复重构对象的缺失度量尺度信息。

算法 2 Monocular 3D reconstruction

1: function TWO – VIEW RECONSTRUCTION

2: Detect features (f_1, f_2) in frames (I_1, I_2)

3: Match f_1 and f_2 between I_1 and I_2

4: Remove false matches → inlier matches $(f^{\wedge}_1, f^{\wedge}_2)$; Geometric Verification

5: 5 – point$(I_1, I_2, f_1, f_2) \to E_s$; Essential Matrix

6: SVD$(E_s) \to R, t$; Relative Camera Pose

7: Projection matrices $P_1, P_2 \to P_1 = [I|0], P_2 = [R|t]$ Triangulate$(I_1, I_2, f_1, f_2, P_1, P_2) \to X^{3D}$; 3D points

8: Bundle Adjustment$(f_1, f_2, P_1, P_2, X^{3D})$

9: end function

10: Import initial pair I_1, I_2

11: TWO – VIEW RECONSTRUCTION(I_1, I_2, X^{3D})

12: for i:1:#*frames* do

13: Import new camera frame I_i

14: PnP$(I_i, X^{3D}) \to P_i$; i_{th} projection matrix, new frame registered

15: Multi – view triangulation → $X^{3D}_{i,k} = X^{3D}, X^{3D}_{i,k}$; Augment existing 3D structure.

16: Bundle adjustment(f_i, P_i, X^{3D}_{new})

17: end for

8.5　数据集收集

8.5.1　试验装置

本章中演示的试验性试验考虑了两种不同的空中平台,图 8.3 所示的 Ascending Technologies NEO(AscTec NEO)六旋翼机和图 8.4 所示的定制四旋翼机。

AscTecNEO 平台直径为 0.59m,高度为 0.24m。每个螺旋桨的长度为 0.28m,如图 8.3 所示。该平台飞行时间为 26min,最大空速可达 15m/s,最大爬升速率可达 8m/s,最大有效载荷可达 2kg。它配有一台板载英特尔 NUC 计算机,配备酷睿 i7-5557U 处理器和 8 GB 内存。NUC 运行安装了机器人操作系统(ROS)①框架的 Ubuntu 服务器 14.04。

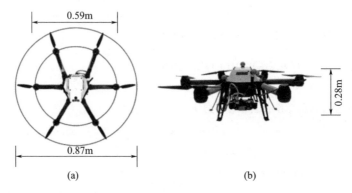

图 8.3 配备视觉传感器的高空作业平台 AscTec NEO

图 8.4 配备视觉传感器的用于地面环境的定制空中平台

定制的低成本 MAV 搭载一块板载 AAEON-UP-board②,配备 64 位英特尔 Atom x5 Z8350 处理器,最高频率可达 1.92GHz,并且搭配 4GB 内存,运行 ROS 框架的 Ubuntu 服务器 16.04。空中平台配备了一个前视摄像头和一根 LED 灯条。飞行控制器是 PIXHAWK[42],平台飞行时间近 20min。

空中检查机上配备的传感器由两种不同的相机组成:① VI 传感器;② FOXEER Box 相机。VI 传感器由 Skybotix AG 制造,质量为 0.12kg(图 8.3),

---

① http://www.ros.org
② https://up-board.org/

放置在空中平台的前部,向地面倾斜45°。该传感器由一个78°FOV的单色全局快门立体相机组成,同时与IMU紧密校正并硬件同步[43]。相机流固定为20帧/s(FPS),分辨率为(752×480)像素。最后,深度范围为0.4~6m。

在隧道试验中,选择了FOXEERBox相机,以60帧/s的速度捕捉帧,分辨率为(1920×1080)像素,对角线FOV为155°。相机套件规格的多样性是出于在具有挑战性的条件下测试性能的需要。

这项研究提出了几个收集视觉数据以检查各种结构的案例。然而,在本研究中使用的数据集中,有两个主要数据集是从限制公众进入的地区收集的。此外,在风力发电场(数据集1)和地下黑暗隧道(数据集2)进行了试验,本节的其余部分将介绍这些数据集。

8.5.2 数据集1

这个数据集是在瑞典Burea的一个风力发电场的多次访问中收集的。在这个位置,测量到风速最高可达13m/s。可视化数据从多个视图中捕捉到一个风力发电机,其底部直径约为4.5m,顶部直径约为1.5m,塔架高度为64m(图8.5)。叶片长度为22m,相应的绳索长度在叶片根部约为2m,叶片顶部约为0.2m,而轮毂和短舱的长度约为4m。访问区域的快照如图8.5所示。

图8.5 MAV检查的风力发电机的访问位置

8.5.3 数据集2

这组数据是在瑞典Lulea的Mjolkudsberget山下的隧道内收集的。选定的环境类似于地下矿山隧道,黑暗,缺乏任何自然照明,而隧道表面由不均匀的岩层组成。试验隧道区域的尺寸为100m×2.5m×3m,MAV沿着隧道的路径飞行的同时捕捉了相机序列。此外,隧道内没有强磁场,飞行过程中有小颗粒飘浮在空中。空中平台配有一根10W的LED灯条,指向相机的FOV,照亮周围环境。更详细地说,这根灯条被设置为不同的照度水平(单位面积光通量或勒克斯(lx)),从2000lx到400lx不等。在数据集序列中使用的相机是FOXEER Box①,它以60帧/s的速度记录了(1920×1080)分辨率的图像。图8.6描述了数据集的代表性快照,显而易见,周围环境是以黑暗为主的。

图8.6 来自瑞典Lulea的Mjolkuddsberget山的隧道的快照,用于数据集收集

8.6 试验结果

本节总结了在三个不同地点进行的现场试验的结果,即在试验室、风力发电场和地下隧道中。结果证明了本章所述方法的适用性,有助于该领域的进一步发展。

8.6.1 室内场景

在这个场景中,一个小物体被放置在FROST试验室的飞行场地内,如图8.7所示。结构尺寸为 $0.57m \times 0.4m \times 0.3m$,已指派两架配备立体相机的代理执行检查任务。在试验室环境下,由VICON运动捕捉(Mo-cap)系统提供代理姿态估计,并将其反馈给飞行控制器和路径规划器。实际结构和重构结构如图8.7所示。

---

① http://foxeer.com/Foxeer-4K-Box-Action-Camera-SuperVision-g-22

第 8 章 无人机航空成像与基础设施重构

(a)　　　　　　　　(b)

图 8.7　待重构的简单室内结构和结构的点云

使用 Autodesk ReCap 360 生成了结构的详细密集三维网格，Autodesk ReCap360① 是一款在线摄影测量软件，适用于精确的三维建模。从图像数据获得的重构表面如图 8.8 所示。

图 8.8　室内结构的协同三维网格模型

8.6.2　室外场景 1

所提出的使用空中平台的三维模型生成系统已额外部署在现实生活的结构体中，以模拟自主视觉检查任务。为此，Lulea 大学的校园喷泉被选为实际基础设施的代表进行空中合作检查。喷泉尺寸为半径 2.8m，高度 10.1m，无额外分支。在这些现场试验中，与之前提出的室内场景相比，运动捕捉系统不可用，因

---

① Autodesk RECAP 360：http://recap360.autodesk.com/

此采用了另一种定位方案。更具体地说，MAV 的定位是基于仅依赖于车载传感器套件的 VI 里程估计。对于重构，从两个空中代理收集的数据集被合并，因为它们覆盖了场景的不同部分，如第 1.4 节所述，并由 SFM 算法①处理。喷泉及其稀疏的三维模型如图 8.9 所示。

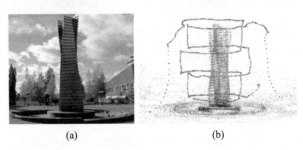

(a)　　　　　　　　　(b)

图 8.9　Lulea 大学室外喷泉和具有估计飞行轨迹的结构的协作点云

它们的起始位置都有一个最大距离，相差 180°。整个飞行时间从 370s 缩短到 189s，平均沿路径的速度为 0.5m/s。图 8.9 中提供的稀疏重构不能用于资产缺陷识别，因为它缺乏纹理信息并且包含噪声。与室内试验类似，从图像数据获得的重构三维网格如图 8.10 所示。

图 8.10　室外结构的协同三维网格

8.6.3　室外场景 2

结果表明，该路径规划器的协同方案可以成功地集成到自动化检查任务中。在这个场景下，飞行器被命令在风力发电机周围导航，并使用高分辨率图像

---

① https://colmap.github.io/

序列和任务后生成的三维模型收集可用于位置感知检查的视觉数据。所提议的检查系统的主要优点是检测到的故障可以与特定的坐标联系起来,这些坐标可以作为维修技术人员的指南,也可以用于随时间推移的临时检查。这是现有空中检查系统的一大贡献,因为它提供了基础设施检查员所需的基本信息,从而实现了安全和自主的空中检查,风力发电场位于瑞典北部(图8.11)。演示的系统在执行导航任务期间处理所有必要的机载处理,同时任务操控员通过 WiFi 链路进行通信。空中平台能够使用惯性-超宽带融合[44-45]状态估计方案围绕大型风力发电机自主飞行,以 IMU(100Hz)的速率运行,而覆盖规划器生成的轨迹在任务开始前已上传到 MAV 处理单元。总的来说,路径是自动跟踪的,没有现场操控员的任何干预,收集的数据保存在机内,下载任务数据后在地面站进行后处理。

图8.11 使用数据集1生成的飞行至20m高度的风力发电机底座的点云

8.6.4 地下场景

本部分考虑的应用场景是利用空中机器人平台检查地下矿山隧道。检查任务的结果将是生成被检查隧道的高保真三维模型。检查专家可以进一步分析重构以检测异常或其他类型的缺陷,从而加快和促进维护任务,甚至对先验挖掘地图库进行整体更新。在这种场景下,MAV 是人工驾驶的,重点是在暗光环境下重构这项具有挑战性的任务。数据集是从位于瑞典 Lulea 的 Mjolkuddsberget 山下隧道内的空中平台收集的。选定的环境类似于地下矿山隧道,没有任何外部照明,漆黑一片,而隧道表面由不均匀的岩层组成。试验隧道区域的尺寸为100m × 2.5m × 3m,拍摄了相机序列,同时 MAV 沿着隧道的路径飞行。空中平台配备了一根 LED 灯条,指向相机的 FOV,照亮周围环境。更详细地说,这根灯条被设置为不同的照度水平,从 4000lx 到 3000lx。图8.12 和图8.13 显示了重构隧道的结果。

图 8.12　使用数据集 2 对访问隧道进行密集重构　　　图 8.13　使用数据集 2 生成的部分访问隧道的三维网格

8.7　未来的发展趋势

MAV 研究不断发展,同时寻求集成技术,这些技术有可能进一步促进其在现实生活应用场景中的建立。目前对微型空中平台的一个限制是机载计算能力。完全自主的检查场景需要对平台进行精确实时的定位、规划、控制和映射,以实现大规模、长期的操作,由于板载的处理能力和内存有限,对当前的解决方案是个挑战。云计算是可以解决这个问题的一个领域,目前许多行业都试图在其运营周期中融入云计算。云计算是一个框架,它通过互联网以用户服务的形式提供必要的资源,如计算能力和基础设施以及应用程序开发,而不受位置或交换数据类型的限制[46]。该框架建立在连接的服务器上,而不是本地计算设备上。这项技术与边缘计算相结合,将允许在平台外分离要求苛刻的计算能力,仅在计算要求有限的情况下保留必要的机载功能。云连接飞行器的新时代将对基础设施检查任务产生重大影响,因为它可以在大规模作业中实时处理数据(如高分辨率视觉数据、海量区域映射数据),提高生产率、有效性和执行任务的速度,同时保持低成本。这成为了空中机器人领域的一个创新引擎和未来趋势,将改变这项技术的游戏规则,并将其确立在最前沿。

随着点云变得越来越密集以及四维系统的不断发展,利用软计算的更好的优化技术将是有必要的[47,48]。

8.8　小结

本章提出了一个利用一个或多个自主 MAV 进行基于图像的三维基础设施

重构的框架，特别关注用于图像采集的传感器选择和用于大规模三维重构的视觉处理方案。在本研究中，代理围绕目标的结构导航，他们通过 SFM 和 MVS 技术收集视频序列进行后处理，目标是生成密集点云。在所提出的方法中，所部署的 MAV 仅依赖于其机载计算机和传感系统，而所提出的框架已经在真实的室内、室外和地下基础设施检查试验中得到了成功的验证。

参考文献

[1] Marins HR, and Estrela VV. On the use of motion vectors for 2D and 3D error concealment in H. 264/AVC video. Feature detectors and motion detection in video processing. IGI Global, Hershey, PA, USA, 2017. p. 164 – 186. doi:10. 4018/978 – 1 – 5225 – 1025 – 3. ch008

[2] Estrela VV, and Coelho AM. State – of – the art motion estimation in the con – text of 3D TV. In Multimedia networking and coding, eds. Reuben A. Farrugia and Carl J. Debono, IGI Global, Hershey, PA, USA, 2013. p. 148 – 173. doi:10. 4018/978 – 1 – 4666 – 2660 – 7. ch006

[3] Kanellakis C, Fresk E, Mansouri SS, and Nikolakopoulos G. Autonomous visual inspection of large – scale infrastructures using aerial robots. arXiv preprint arXiv:190105510. 2019.

[4] Kanellakis C, Mansouri SS, Georgoulas G, et al. Towards autonomous survey – ing of underground mine using MAVs. In: International Conference on Robotics in Alpe – Adria Danube Region. Springer; Zurich, Switzerland, 2018. p. 173 – 180.

[5] Kanellakis C, and Nikolakopoulos G. Survey on computer vision for UAVs: current developments and trends. Journal of Intelligent & Robotic Systems. 2017;1 – 28.

[6] SESAR Joint Undertaking. European drones outlook study. Unlocking the value for Europe. 2016.

[7] Mansouri SS, Kanellakis C, Fresk E, Kominiak D, and Nikolakopoulos G. Cooperative coverage path planning for visual inspection. Control Engineering Practice. 2018;74:118 – 131.

[8] Lee AC, Dahan M, Weinert AJ, and Amin S. Leveraging sUAS for infra – structure network exploration and failure isolation. Journal of Intelligent & Robotic Systems. 2019;1 – 29.

[9] Yuan C, Liu Z, and Zhang Y. Learning – based smoke detection for unmanned aerial vehicles applied to forest fire surveillance. Journal of Intelligent & Robotic Systems. 2018. Available from https://doi. org/10. 1007/s10846 – 018 – 0803 – y.

[10] Sampedro C, Rodriguez – Ramos A, Bavle H, Carrio A, de la Puente P, and Campoy P. A fully – autonomous aerial robot for search and rescue applica – tions in indoor environments using learning – based techniques. Journal of Intelligent & Robotic Systems. 2018. Available from https://doi. org/10. 1007/s10846 – 018 – 0898 – 1.

[11] Morgenthal G, and Hallermann N. Quality assessment of unmanned aerial vehicle (UAV) based visual inspection of structures. Advances in Structural Engineering. 2014;17(3):289 – 302.

[12] Jordan S, Moore J, Hovet SE, et al. State – of – the – art technologies for UAV inspections. IET

Radar, Sonar & Navigation. 2017;12(2):151 – 164.

[13] Achtelik MW, Weiss S, Chli M, Dellaert F, and Siegwart R Collaborative stereo. In: 2011 IEEE/RSJ International Conference on Intelligent Robots and Systems (IROS). IEEE; San Francisco, California, USA, 2011. p. 2242 – 2248.

[14] Perez – Grau FJ, Ragel R, Caballero F, Viguria A, and Ollero A. An architecture for robust UAV navigation in GPS – denied areas. Journal of Field Robotics. 2018;35(1):121 – 145.

[15] Achtelik MW, Lynen S, Weiss S, Chli M, and Siegwart R. Motion – and uncertainty – aware path planning for micro aerial vehicles. Journal of Field Robotics. 2014;31(4):676 – 698.

[16] Scaramuzza D, Achtelik MC, Doitsidis L, et al. Vision – controlled micro flying robots: from system design to autonomous navigation and mapping in GPS – denied environments. IEEE Robotics & Automation Magazine. 2014; 21(3):26 – 40.

[17] Lupashin S, Hehn M, Mueller MW, Schoellig AP, Sherback M, and D'Andrea R. A platform for aerial robotics research and demonstration: the flying machine arena. Mechatronics. 2014; 24(1):41 – 54.

[18] Teixeira L, and Chli M. Real – time local 3D reconstruction for aerial inspection using superpixel expansion. In: 2017 IEEE International Conference on Robotics and Automation (ICRA). IEEE; Singapore, 2017. p. 4560 – 4567.

[19] Forster C, Faessler M, Fontana F, Werlberger M, and Scaramuzza D. Continuous on – board monocularvision – based elevation mapping applied to autonomous landing of micro aerial vehicles. Imaging and Sensing for Unmanned Aerial Vehicles: Volume 2 – Deployment and Applications aerial vehicles. In: 2015 IEEE International Conference on Robotics and Automation (ICRA). Seattle, WA, USA, IEEE; 2015. p. 111 – 118.

[20] Schafer BE, Picchi D, Engelhardt T, and Abel D. Multicopter unmanned aerial vehicle for automated inspection of wind turbines. In: 2016 24th Mediterranean Conference on Control and Automation (MED). IEEE; Athens, Greece, 2016. p. 244 – 249.

[21] Zhang C, and Elaksher A. An unmanned aerial vehicle – based imaging system for 3D measurement of unpaved road surface distresses 1. Computer Aided Civil and Infrastructure Engineering. 2012;27(2):118 – 129.

[22] Mancini F, Dubbini M, Gattelli M, Stecchi F, Fabbri S, and Gabbianelli G. Using unmanned aerial vehicles (UAV) for high – resolution reconstruction of topography: the structure from motion approach on coastal environments. Remote Sensing. 2013;5(12):6880 – 6898.

[23] Schmid K, Lutz P, Tomic T, Mair E, and Hirschmüller H. Autonomous vision – based micro air vehicle for indoor and outdoor navigation. Journal of Field Robotics. 2014;31(4):537 – 570.

[24] Gohl P, Burri M, Omari S, et al. Towards autonomous mine inspection. In: 2014 3rd International Conference on Applied Robotics for the Power Industry (CARPI). IEEE; Foz do Iguassu, Brazil, 2014. p. 1 – 6.

[25] Ozaslan T, Loianno G, Keller J, et al. Autonomous navigation and mapping · for inspection of penstocks and tunnels with MAVs. IEEE Robotics and Automation Letters. 2017;2(3):1740 – 1747.

[26] Bakambu JN. Integrated autonomous system for exploration and navigation in underground mines. In: 2006 IEEE/RSJ International Conference on Intelligent Robots and Systems. IEEE; Beijing, China, 2006. p. 2308 – 2313.

[27] Losch R, Grehl S, Donner M, Buhl C, and Jung B. Design of an autonomous robot for mapping, navigation, and manipulation in underground mines. In: 2018 IEEE/RSJ International Conference on Intelligent Robots and Systems (IROS). IEEE; Madrid, Spain, 2018. p. 1407 – 1412.

[28] Tardioli D, Sicignano D, Riazuelo L, Romeo A, Villarroel JL, and Montano L. Robot teams for intervention in confined and structured environments. Journal of Field Robotics. 2016; 33(6): 765 – 801.

[29] Estrela VV, Saotome O, Loschi HJ, et al. Emergency response cyber – physical framework for landslide avoidance with sustainable electronics. Technologies. 2018; 6: 42. doi: 10.3390/technologies6020042.

[30] Zlot R, and Bosse M. Efficient large – scale three – dimensional mobile mapping for underground mines. Journal of Field Robotics. 2014; 31(5): 758 – 779.

[31] Wang W, Dong W, Su Y, Wu D, and Du Z. Development of search – and – rescue robots for underground coal mine applications. Journal of Field Robotics. 2014; 31(3): 386 – 407.

[32] Tardioli D, Riazuelo L, Sicignano D, et al. Ground robotics in tunnels: Keys and lessons learned after 10 years of research and experiments. Journal of Field Robotics. 2019; 36: 1074 – 1101.

[33] Endres F, Hess J, Engelhard N, Jurgen S, Cremers D, and Burgard W. An evaluation of the RGB – D SLAM system. In: 2012 IEEE International Conference on Robotics and Automation (ICRA). IEEE; Saint Paul, Minnesota, USA, 2012. p. 1691 – 1696.

[34] Besl PJ, and McKay ND. Method for registration of 3 – D shapes. In: International Society for Optics and Photonics Robotics – DL tentative. San Diego, CA, USA, 1992. p. 586 – 606.

[35] Labbe M, and Michaud F. Memory management for real – time appearance – based loop closure detection. In: 2011 IEEE/RSJ International Conference on Intelligent Robots and Systems. IEEE; San Francisco, California, USA, 2011. p. 1271 – 1276.

[36] Sivic J, and Zisserman A. Efficient visual search of videos cast as text retrieval. IEEE Transactions on Pattern Analysis and Machine Intelligence. 2009; 31(4): 591 – 606.

[37] Schonberger JL, and Frahm JM. Structure – from – motion revisited. In: Proceedings of the IEEE Conference on Computer Vision and Pattern Recognition; Las Vegas, Nevada, USA, 2016. p. 4104 – 4113.

[38] Wu C. Towards linear – time incremental structure from motion. In: 2013 International Conference on 3D Vision – 3DV 2013. IEEE; Seattle, Washington, USA, 2013. p. 127 – 134.

[39] Hartley R, and Zisserman A. Multiple view geometry in computer vision. Cambridge: Cambridge University Press; 2003.

[40] Nister D. An efficient solution to the five – point relative pose problem. IEEE Transactions on Pattern Analysis and Machine Intelligence. 2004; 26(6): 756 – 770.

[41] Gao XS, Hou XR, Tang J, and Cheng HF. Complete solution classification for the perspective – three – point problem. IEEE Transactions on Pattern Analysis and Machine Intelligence. 2003;

25(8):930 – 943.

[42] Meier L, Tanskanen P, Heng L, Lee GH, Fraundorfer F, and Pollefeys M. PIXHAWK: a micro aerial vehicle design for autonomous flight using onboard computer vision. Autonomous Robots. 2012;33(1 – 2):21 – 39.

[43] Fresk E, Mansouri SS, Kanellakis C, Halen E, and Nikolakopoulos G. Reduced complexity calibration of MEMS IMUs. In: 25th Mediterranean Conference on Control and Automation (MED); Valletta, Malta, 2017.

[44] Fresk E, Odmark K, and Nikolakopoulos G. Ultra wideband enabled inertial odometry for generic localization. IFAC – PapersOnLine. 2017;50(1): 11465 – 11472.

[45] Arshaghi A, Razmjooy N, Estrela VV, et al. Image transmission in MIMO – OSTBC system over Rayleigh channel using multiple description coding (MDC) with QPSK modulation. In: Estrela VV, Hemanth J, Saotome O, Nikolakopoulos G, and Sabatini R (Eds), Imaging and Sensing for Unmanned Aircraft Systems, Vol. 2, IET, 2020.

[46] Kushida KE, Murray J, and Zysman J. Diffusing the cloud: cloud computing and implications for public policy. Journal of Industry, Competition and Trade. 2011;11(3):209 – 237.

[47] Razmjooy N, and Estrela VV. Applications of image processing and soft computing systems in agriculture. IGI Global; Hershey, PA, USA, 2019. p. 1 – 337. doi: 10.4018/978 – 1 – 5225 – 8027 – 0.

[48] Hemanth J, and Estrela VV. Deep learning for image processing applica – tions. Advances in parallel computing. IOS Press; Netherlands, 2017. ISBN 978 – 1 – 61499 – 821 – 1 (print) 978 – 1 – 61499 – 822 – 8 (online).

第9章 深度学习作为无人机系统中超分辨率成像的可替代方案

本章提出了一个超分辨无人机捕获低分辨率(LR)图像的架构。该架构用一个卷积神经网络超分辨低分辨率图像。该架构还去除了低分辨率图像中的模糊。提出的系统用峰值信噪比、像素域的结构相似性(SSIM)和视觉信息保真度(VIFP)进行评估。试验结果验证了与其他基于定性和定量分析的最先进算法相比所提出的方法的优势。本章末尾讨论了无人机超分辨率(SR)成像的未来趋势,最后是结论部分。

9.1 引言

近些年,无人机的应用逐步增长[1-3]。无人机是一种可以手动遥控,或在预编程 GPS 计划航路点飞行,或在更复杂的机载智能系统引导下自主飞行的飞机。最近,无人机已广泛应用于民用和军用,特别是在情报、监视和侦察,以及遥感中。实践中,由于成像系统和大气条件欠佳的限制,小型无人机捕获的监视图像几乎都是模糊或退化的。

就室内外环境监视和遥感而言,无人机信息-物理系统(UAV-CPS)非常有用[4]。多模态传感器采集的航空图像的配准/融合/拼接和压缩需要创新的算法实现无中断的更精确的局部化和映射。更精确的系统和算法可支持并提高遥感操作。目前,GPS 还用于无人机发现其他无人机和被监测物体的坐标。此外,众所周知 GPS 有覆盖范围和精度问题。

由于风的作用,小型无人机通常有振动。这样,缺乏稳定的视频/图像很难瞄准关注的目标[5-8]。创建无人机图像方面的主要问题是捕获的视频或图像质量和飞行时间限制。获得大区域机载图像需要很长时间。设置无人机最佳飞行高度减少了所用时间。然而,增加高度还扩大了无人机的视野,从而潜在地缩短了飞行时间,并减少目标物体的视觉细节。因此,必须使用超分辨率程序实现目标物体的高分辨率(HR)、高精度图像。超分辨率图像技术能够通过不同帧的图像配准和融合,从一组低分辨率图像产生一幅直接与小型载荷无人机获得的监视视频相联系的高分辨率图像。

超分辨率方法可以是双重的:即多帧(MF)和单帧(SF)[9]。基于单帧的超分辨率方法用单幅图像(SI)生成对应的高分辨率版本图像[5-7],可采用插值(最近邻、双线性、双三次等)和基于举例的方法[10-13]。由于缺乏输入数据量(图像),单帧图像超分辨率(SISR)呈现不良的结果。例如,插值法产生模糊和失真图像。此外,基于举例的方法计算无效,可能难以使用或不适合实时处理。关于单帧超分辨率(SFSR)的详细讨论不在本章范围内[6,14-20]。多帧超分辨率(MFSR)法通过融合来自一个场景被破坏的几幅图像[21-24]的一些信息片段,产生一幅高分辨率图像。由于相机噪声周替、转换或仿射变换,如相机摇动、不同的捕获和曝光时间和相机和场景间的相对位移,类似的图像预期不同。这一额外的知识有助于通过在一个基准高分辨率栅格上排列这些帧,估计这些特征,提高场景分辨率。

9.2节讨论了一个依赖卷积神经网络(CNN)和去模糊(dehazing)超分辨率模型;9.3节介绍将用作案例研究的试验;9.4节分析涉及超分辨率在UAV-CPS中使用和实现的关键问题;9.5节给出结论。

9.2 超分辨率模型

超分辨率方案用低分辨率图像重构高分辨率图像。在超分辨率技术中,图像捕获过程中必须恢复丢失的高频内容。从下采样低分辨率观察,超分辨率算法主要关切重构高分辨率图像;它从模糊、有噪点和退化的图像产生高质量图像。克服低分辨率成像系统固有分辨率限制的技术特性突现,相当于词语'超级'。超分辨率方法的优点是:① 成本低;② 可使用现有的低分辨率成像系统,无任何额外的硬件;③ 具有灵活性。

图像复原问题与超分辨率重构问题紧密相连。图像复原的目的是从退化(如噪声和模糊)中恢复图像。因此,对于图像复原问题,存储图像和观测图像大小相同。超分辨率包括不同的高分辨率和低分辨率图像尺寸,因此有对超分辨图像采样因子的依赖。

这里需要理解细节,通过捕获图像及其进行的超分辨率算法变换。传统的数字图像捕获系统(相机)中通常有聚焦镜头、处理器芯片、光学传感器、电子电路及其他机械子系统[25-31]。当用这样一台相机捕获一个场景时,高分辨率图像经历序列退化,例如:模糊、附加噪声和下采样。由于光学失真(模糊、衍射极限和传感器密度不足),有空间分辨率的自然损失。由于导致相机和物体之间的相对运动、光学像差、限定快门速度和大气湍流作用,因此观测会模糊。图像还会受传输过程中或传感器中出现的各类噪声影响而退化。由于相机运动(如变焦、倾斜和平移)用视频相机获得的图像帧可以旋转和标定。这样,由于观测物之间的相对运动会引起模糊,因此观测图像说明了高分辨率图像的退化。一个代表图像捕获过程的数学模型明确表达分析了超分辨率重构问题。

观测低分辨率图像表现的原始高分辨率图像由该模型建立联系,称为正向模型或观测[32-34]。任何成功的超分辨率方法中的关键是正确的建立观测模型,构想中转换、模糊、失真和噪声是最常用的超分辨率重构正向模型。图9.1 给出了典型的超分辨率模型。

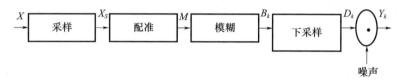

图 9.1 超分辨率模型

图中:X 表示连续场景;X_S 表示从限带连续场景中以高于奈奎斯特(Nyquist)速度采样的理想高分辨率图像;输出 Y_k 是图像传感器观察的第 k 个低分辨率图像,观测模型由下式表示:

$$Y_k = DB_kM_kX + N_k \qquad k = 1,2,3,\cdots,K \qquad (9.1)$$

式中:D 表示下采样算子;B_k 包含第 k 个低分辨率图像的模糊表示;M_k 包含把第 k 个低分辨率图像转换成高分辨率图像栅格的运动信息;N_k 是第 k 个低分辨率图像中的噪声。

深度学习(DL)法已成功应用于图像复原问题。基于深度神经网络模型,提出了用于图像超分辨率的各种方法,得到比其他模型更好的性能。基于深度卷积神经网络的单帧图像超分辨率方法出现在参考文献[35]中,其研究了低分辨率和高分辨率图像间的连续映射,这里,基于粗编码的超分辨率方法可被视为一个深度卷积网络观察。该网络结构简单,可用于实时应用处理速度快的场景。我们研究了许多网络结构和参数设置,并分析了性能和速度。并将本章延伸到 RGB 色彩通道管理,且表现出更好的重构质量性能;另一个随梯度转换展开的单帧图像超分辨率深度学习方法出现在参考文献[36]中。

一个卷积神经网络对低分辨率图像采样,计算放大的(up-scaled)图像梯度,并用梯度变换网络将其转换成所需要的梯度,建立重构的能量函数。

最后,优化能量函数以估计高分辨率图像,该方法产生一幅清晰的高分辨率图像及振铃假象。

参考文献[37]提出了一个用深度卷积神经网络的光场图像超分辨率方法。

提出的方法分析了现实的应用,如重新调焦和深度图估计。参考文献[38]报道了可设计为一个神经网络,并在级联结构中端对端训练的粗编码模型。Kato 和 Ohtani[39]利用多通道卷积神经网络提出了一个新颖的超分辨率方法。

在该方法中,根据输出位置把像素分成四组。这四组像素被创建为一个 2 × 2 的放大图像。这种架构独立于双三次插值法的使用。与传统的超分辨率卷积神经网络(图 9.2)相比,用该方法平均信噪比峰值(PSNR)高于 0.39dB,提出的方法由一些后续块组成。

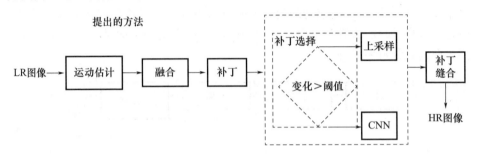

图 9.2　所提出方法的体系结构

9.2.1　运动估计

输入的低分辨率图像间存在相对漂移的可能性。识别最佳质量的无人机图像后,最好在超分辨过程开始前全面对准所有其他输入的低分辨率帧。通过全面校正低分辨率图像,可以使局部斑块域错误校正的可能性最小化,从而提高局部校正的成功率。校正过程中选择的最佳帧被用作一个模板图像。为发现模板图像和低分辨率图像间的相对漂移,参考文献[40]提出改进菱形搜索算法,使用并讨论如下。

提出的改进菱形搜索算法是对十字菱形搜索的改进[41],以适应图像的小中心偏差特性。图像被分成为(16 × 16)像素的宏块。宏块中 4 个(上、下、左、右)搜索图案如图 9.3 所示。

算法

(1)第 1 步:大菱形搜索图案(SP)位于搜索窗口中心。计算 9 个点的绝对误差之和(SAE),确定最小绝对误差点。绝对误差用下式计算:

$$SAE = \sum_{i=0}^{P-1} \sum_{j=0}^{Q-1} |C_{ij} - R_{ij}| \qquad (9.2)$$

式中:$P \times Q$ 为宏块大小;C_{ij} 和 R_{ij} 分别是在当前宏块和参考宏块中比较的像素,使绝对误差最小化的宏块将变成在那个位置块的运动向量。

(2)第 2 步:如果最小点不在原点,则修正原点并进行搜索。如果最小绝对误差出现在中心,则搜索停止,另外修正中心并进一步处理。

第 9 章 深度学习作为无人机系统中超分辨率成像的可替代方案

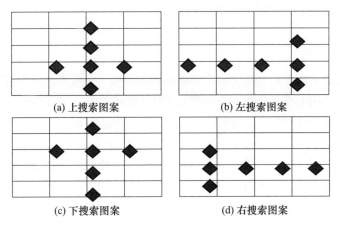

图 9.3 宏块中的搜索图案
(◆代表在一个宏块中的搜索位置)

(3)第 3 步:基于从第 2 步得到的最小点,选择图 9.3 对应的图案。这里,搜索点数减少到 6 个。若前一步的最小点出现在搜索图案的上顶点,则新的搜索图案为图 9.3(a)。

① 若前一步的最小点出现在搜索图案的下顶点,则新的搜索图案为图 9.3(b)。

② 若前一步的最小点出现在搜索图案的左顶点,则新的搜索图案为图 9.3(c)。

③ 若前一步的最小点出现在搜索图案的右顶点,则新的搜索图案为图 9.3(d)。

④ 若最小绝对误差出现在搜索图案的中心,则搜索停止,另外修正中心,并做进一步处理。最小绝对误差点与必要的运动向量相符。

这些运动向量告诉我们为何图像中许多像素应当移位,因为它与另一图像对准。在校正阶段,每个输入的低分辨率图像随精确平移变化而变化,对于无人机模板图像,沿相反方向把所有图像与模板图像对准。

无人机超分辨成像并不是使用全局超分辨率算法,而是使用基于局部斑块的超分辨率方法。在该方法中,整个图像被分成正方形区的子集。局部变形可用局部斑块避免,在极域无人机图像上完成超分辨率。经帧校正后,每一帧被分成 10×10 大小的斑块,属于相同位置不同图像的斑块被存储在相同的局部集中。根据它们的位置,图像斑块被重新整理和聚合。"局部集"中的图像斑块随斑块的一级导数增加,由 Sobel 滤波器产生,以得到一个边缘增强的斑块。下一步是融合多个斑块的像素信息。

本章中,一项基于偏差、在离散余弦变换(DCT)域计算的图像融合技术[41]应用于融合无人机图像。计算(8×8)像素图像块的离散余弦变换系数。每一块的偏差决定整块的质量。选择最佳质量块与图像中的另一块融合。这样,融合来自多个局部斑块的像素密度信息,并产生一个最终的斑块。该斑块应该清晰、干净和明显。斑块被送至超分辨率系统。

9.2.2 去模糊

无人机捕获的图像,包括造成非清晰图像的模糊。去模糊块减少捕获图像中的模糊量[42-43]。用一个暗通道先验(DCP)法降低模糊[44-45]。基于室外无模糊图像统计,该方法通常用于产生一幅自然的无模糊图像。模糊图像 $S(x)$ 表示为

$$S(x) = Z(x)t(x) + A[1 - t(x)] \tag{9.3}$$

式中:A 为大气光;$Z(x)$ 为前景密度;$t(x)$ 为残留能量的百分比。

本章提出的方法用 DCP[27] 表示为

$$D^x = \min_{y \in \Omega(x)} \left(\min_{c \in \{r,g,b\}} D^C(y) \right) \tag{9.4}$$

式中:$D^C(y)$ 是 D 的色彩通道;$\Omega(x)$ 是一个以 x 为中心的局部斑块;$\min_{y \in \Omega(x)}$ 是一个最小滤波器。

介质传输通过下式估计:

$$\overset{\vee}{t} = 1 - \min_{y \in \Omega(x)} \left(W \min_{c \in \{r,g,b\}} \frac{D^C(y)}{A^C} \right) \tag{9.5}$$

式中:A^C 为大气光;W 为图像中区别图像深度的少量模糊。

场景辐射率 $J(x)$ 用大气光和传输介质恢复。$J(x)$ 可用下式计算:

$$J(x) = \frac{S(x) - A^C}{\max(t(x), th)} + A^C \tag{9.6}$$

式中:th 是一个阈值,避免分母低值。

9.2.3 斑块选择

并非使用全局超分辨率算法,基于局部斑块的超分辨率方法被用于超分辨无人机图像。在该方法中,整个图像被分成若干个正方块。

通过计算每个斑块的偏差,分析该块中无人机图像斑块的内容。如果斑块的偏差超过预定值,那么超分辨率斑块被送至卷积神经网络块。如果偏差小于预定值,那么斑块被送至上采样块。如果卷积神经网络处理包含少量信息的斑块,那么总体处理时间将增加。为了避免这种情况,包含少量信息的斑块经上采样处理。这种斑块选择方法提高了超分辨率处理速度。

9.2.4 超分辨率

该图像是用如参考文献[35]中讨论的一个卷积神经网络超分辨的,卷积神经网络概况如图9.4所示。

无人机图像的超分辨率通过下列步骤实现。

(1)第一层:该层称为斑块提取和表示层。常见的图像重构策略是通过提取斑块,并由一组预先训练的基数代表。该方法类似于由一组滤波器卷积的图像,每一个均在该层中,图像重叠斑块被提取,并由一组预先训练的离散余弦变换基数表示。设 y 为低分辨率无人机图像,上采样到所需的尺寸(用双三次插值法得到 Y)。目的是从 Y 和图像 $F(x)$ 恢复,尽可能与人工标注框高分辨率图像 Y 相似。该层可表示为

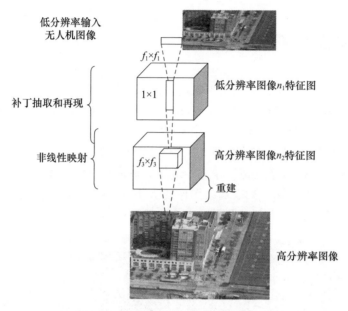

图9.4 利用卷积神经网络的超分辨率

$$F_1(Y) = \max(0, W_1 * Y + \boldsymbol{B}_1) \tag{9.7}$$

式中: W_1 是一个尺寸为 $m \times f_1 \times f_1 \times n_1$ 的卷积滤波器, f_1 为滤波器的空间大小, n_1 为滤波器数, m 为通道数; $\boldsymbol{B}_1$ 是一个偏差。输出由 n_1 特征图组成。$\boldsymbol{B}_1$ 是一个 n_1 维向量,其每个点与一个滤波器相联系。

(2)第二层:该层称为非线性映射层。n_1 维特征向量被映射到 n_2 维特征向量。该层通过下式实现:

$$F_2(Y) = \max(0, W_2 * F_1(Y) + \boldsymbol{B}_2) \tag{9.8}$$

式中: W_2 的大小为 $n_1 \times 1 \times 1 \times n_2$; $\boldsymbol{B}_2$ 为 n_2 维向量, n_2 为特征图数。

(3)第三层:该层称为重构层。所有预测的高分辨率斑块经平均产生最终的高分辨率图像。该层通过下式实现:

$$F(Y) = W_3 * F_2(Y) + \boldsymbol{B}_3 \tag{9.9}$$

式中: W_3 的尺寸为 $n_2 \times f_3 \times f_3 \times k$; $\boldsymbol{B}_3$ 是一个 l 维向量。

在训练过程中,必须确定网络参数 $P = (W_1, W_2, W_3, \boldsymbol{B}_1, \boldsymbol{B}_2, \boldsymbol{B}_3)$ 以学习端对

端映射函数 F。这些网络参数通过减小高分辨率图像 X 和重构图像 $F(Y,P)$ 之间的损失得到。假设 Y_i 为低分辨率图像，Y_i 为对应的高分辨率图像。损失函数用下式计算：

$$L(P) = \frac{1}{n}\sum_{i=1}^{n} \| F(Y_i,P) - X_i \|^2 \tag{9.10}$$

随机梯度下降与标准后向传播[46]被用于使损失最小。

9.3　试验和结果

本章所提出的方法的试验结果通过分析提出的超分辨率系统的性能得到演示。算法用 MATLAB2012 在英特尔 Core i5 机器上实施，RAM 空间为 8 GB。

本章所提出的技术的效能通过无人机图像试验验证。

训练过程中，使用了 220 个训练图像。这些图像被细分成(64×64)像素的子图像，得到约 28000 个图像。在这些试验中滤波器的尺寸被设为 $f_1 = 9$ 和 $f_3 = 5$。特征图 $n_1 = 64$ 和 $n_2 = 32$。输入图像的大小为(640×480)像素，上采样因子为 2。图 9.5 给出了采样输入图像。

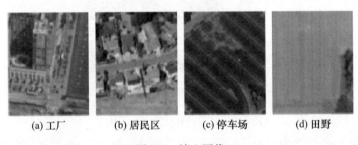

(a) 工厂　　(b) 居民区　　(c) 停车场　　(d) 田野

图 9.5　输入图像

图像质量评估(IQA)法帮助评估系统的性能[47-52]，如 SSIM、VIFP 域、PSNR 和计算时间。SSIM 是一种受对结构成分失真敏感的人类视觉系统信息激发的质量评估方法。该方法评估两个图像间出现的结构成分变化。两个图像间的 SSIM 计算如下。把原始图像和超分辨的图像分成一些斑块。相似性 $S(x,y)$ 可用下式计算：

$$S(x,y) = \left(\frac{2\mu_x\mu_y + K_1}{\mu_x^2 + \mu_y^2 + K_1}\right)\left(\frac{2\sigma_x\sigma_y + K_2}{\sigma_x^2 + \sigma_y^2 + K_2}\right)\left(\frac{2\sigma_{xy} + K_3}{\sigma_x\sigma_y + K_3}\right) \tag{9.11}$$

式中：μ_x 和 μ_y 代表斑块 x 和 y 的平均；σ_x 和 σ_y 表示斑块 x 和 y 的标准偏差。K_1、K_2、K_3 是防止统计不稳定(当被分母所除时发生)的正常数；σ_{xy} 为 x 和 y 的交叉相关：

$$\sigma_{xy} = \frac{1}{\Delta_v - 1}\sum_{i=1}^{T}(x_i - \mu_x)(y_i - \mu_y) \tag{9.12}$$

式中：x_i 和 y_i 分别为图像中斑块 x 和 y 的像素密度；T 为每个斑块 x 和 y 中的像素数。

SSIM 图像评价通过取图像中斑块 SSIM 的平均值得到。SSIM 值的范围为 0~1，为了获得更好的图像质量，SSIM 值应由更大的值确定。VIFP 是无失真时第一阶段和最后阶段视觉通道和失真大块输入和视觉系统大块输出公共数据的统计。成果是一个保真度措施。

9.3.1 峰值信噪比

峰值信噪比（PSNR）为图像和噪声功率中最高可能像素值之间的比率，根据对数分贝（dB）表示，由下式计算出：

$$\text{PSNR}_{\text{dB}} = 10\lg \frac{255^2}{\text{MSE}} \tag{9.13}$$

均方差（MSE）可表示为

$$\text{MSE} = \frac{1}{M\Delta rN} \sum_{i=0}^{M-1} \sum_{j=0}^{N-1} (C_{ij} - R_{ij})^2 \tag{9.14}$$

式中：M 和 N 分别为图像的高和宽；C_{ij} 为原始图像；R_{ij} 是一个失真图像。

根据图 9.6，提出的方法给出比最先进的算法更好的图像质量。为进一步分析所提出的方法，上采样因子设为 4 和 6。提出的方法对上采样因子 4 和 6 的性能如表 9.1 所示。

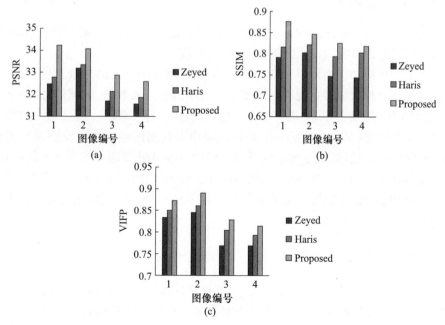

图 9.6 提出的方法用于无人机图像上采样因子 2 的性能分析
(a) PSNR；(b) SSIM；(c) VIFP。

表9.1 上采样因子2和4的性能分析

图像编号		Zeyed			Haris			本章提出的方案		
		PSNR,/dB	SSIM	VIFP	PSNR,/dB	SSIM	VIFP	PSNR,/dB	SSIM	VIFP
上采样因子4	1	31.88	0.759	0.764	32.29	0.787	0.798	33.47	0.808	0.849
	2	32.34	0.767	0.783	32.86	0.779	0.801	33.5	0.811	0.86
	3	31.17	0.696	0.721	31.63	0.742	0.768	31.88	0.783	0.812
	4	30.85	0.704	0.743	31.05	0.722	0.751	31.29	0.791	0.79
上采样因子6	1	30.44	0.712	0.732	31.54	0.729	0.755	31.83	0.793	0.799
	2	31.26	0.705	0.755	31.73	0.726	0.782	32.27	0.756	0.805
	3	29.07	0.649	0.675	30.15	0.669	0.695	30.87	0.711	0.773
	4	28.93	0.673	0.691	29.43	0.692	0.711	30.19	0.721	0.759

9.4 UAV-CPS中超分辨率运用的关键问题

下面讨论一些很快出现的、具有前景的研究方向。

9.4.1 大数据

超分辨率用于遥感图像的关键挑战是时间差引起的室外场景改变,同时使目前的程序适合日常观测大数据(BD)的数据量。

许多大数据知识库可以公开获取,特别是在遥感中。借助海量全局数据,瓶颈问题是把原始数据转化为有用的知识和表示,例如,元数据、地理信息、语义内容、几何事实等,以更好地、几乎全自动地处理它们,合作观测并连接相关的独立程序。机器学习(ML)解开了一个处理和利用大数据的新方法。特别是对于包括多时间数据的使用,重复神经网络(RNN)可外在地处理动态行为和时间序列。这对填补获取数据空白有极大帮助,如由于云或不完整的历史覆盖。此外,它推动了事件预报,如气象观测、农作物估产。在UAV-CPS中积极使用超分辨率参数的有效估计有重要意义。

9.4.2 云计算服务

超分辨率需要解决不适定反转问题(需要极其繁重的计算负担)。在像UAV-CPS的实际应用中,将不得不减少这样花费高昂地消耗计算资源,同时为用云/边缘/极限计算以及智能计算技术抑制最优化所需运算数的自适应滤波和实时策略留下空间。

另一问题是必须融合几种成像模态,以提供更好的图像细节和描述,它要求密集使用云/雾/极限计算和无人机载和非机载通信。单个用户大量数据集的能力有限。最近,许多大的国际技术公司为用户和科学家提供支持拍字节(petabyte)规模的地理空间数据科学分析的工具。谷歌 Earth 搜索引擎是一个在线存储和组织卫星图像的平台,同时还提供便利的工具浏览和搜索数据集。利用这些工具,促进了全球数据挖矿,算法可广泛应用于大量可用数据,无须在用户方下载或存储任何数据。所有计算在云端进行,结果几乎即刻呈现给用户。

此外,数据采集包括所有"哨兵"2 号和 Landsat 采集,包括许多 MODIS 数据、降水数据和高程数据等。

TensorFlow 环境一旦连接平台后,谷歌 Earth 搜索引擎不久也将能够运行现代化的深度学习模型。TensorFlow 是一个提供高效深度学习工具的计算机程序库。除谷歌外,亚马逊也提供云计算服务、Amazon Web 服务。类似地,该平台包含开放的地理空间数据,包括 Landsat 8 和"哨兵"2 号资料库。各种插件程序和服务使用该平台。通常,除了这些为地理空间数据分析和直观化而建的平台外,云服务还能够大规模计算,包括深度学习,云计算正在不断取得进展。可以期待,最终将带来尚待发现的新应用。

9.4.3 图像捕获硬件限制

在可能的情况下,由于错误图像配准所致的误差应被视为模型的一部分,并依赖非高斯附加噪声[53]。

更多的色彩通道交互工作及获得反射率变化的方法可提高图像配准效果。

对于一个或多个低分辨率图像正在失去的情况,将对关联的最优化问题努力产生影响。传感器故障和缺点可以以模型表示,容忍与高分辨率图像有关的一些控制和容错机制。

如果像以前所做的案例研究一样引入模糊,那么可以设计更好的盲方法处理模糊源,同时产生高分辨率图像[54]。有关信噪比、照射变化、阻塞和获得的低分辨率图像模糊的额外信息,可帮助平衡得到高分辨率图像所涉及的计算努力。

高光谱图像(HSI)的使用正在增加。这挑战超分辨率高光谱图像,并且也依赖特殊的传感器配置及全色图像(三频段 RGB 图像)。此外,在设计步骤增加超分辨率并方便地适应所有 UAV-CPS 特性可能是值得的。优化传感器安排的研究将允许通过利用融合所有频段的信号本性,确保更好的成果[55]。

高光谱图像相机可获得并实时完成图像计算任务,导致来自高冗余四维(4D)Cubes 的额外信息,以允许完全的时序动态研究。多时未混合处理可能从多帧超分辨率技术知识中受益。快速、低反应时间高光谱图像分析鼓励把非昂贵的传感器用在小型智能设备上。低分辨率高光谱图像的超分辨率与通常的高分辨率 RGB 智能手机相机可提供实时分析。

有损耗的 GPU-native 压缩格式,如 BCn、ETC 和 ASTC 已被广泛应用于减少图像存储器纹理痕迹。类似一款手机 Mobile 3D 游戏,可下载的 GPU 应用程序应选择最适当的压缩格式和每个纹理的分辨率,同时考虑多种目标设备的能力和下载量[55-58]。

高光谱图像提供丰富的光谱特征,并有助于多种视觉任务[59-62]。然而,由于硬件限制,不仅采集高质量高光谱图像比采集 PAN 更困难,而且采集的高光谱图像的分辨率更低。这样,将超分辨率引入该领域,研究人员倾向于综合高分辨率 PAN 和低分辨率高光谱图像预测高分辨率高光谱图像。在他们当中,Huang 等[63]介绍了一个学习低分辨率-高分辨率映射与 PAN,并将其传递到高光谱图像的稀疏去噪的自动编码器。Masi 等[64]使用 SRCNN[65]并结合几幅提升性能的非线性无线电系数的图。Wei 等[66]提出了一个基于残留学习的更深层 DRPNN[67],并实现更高的空间-光谱统一的精度。最近,Qu 等[68]联合训练两个编码器-解码器网络,分别在 PAN 和高光谱图像上完成高分辨率,并由共享解码器把 PAN 域超分辨率知识传递到高光谱图像域,并应用约束,如角相似性损失和重构损失。

9.4.4 视频超分辨率

根据视频超分辨率,多帧提供更多的场景信息,不仅有帧内空间依赖,而且有帧间时间依赖(如运动、亮度和色彩变化)。这样,现有的工作主要聚焦于更好使用空时依赖,包括外在的运动补偿(如光流算法[69-71],基于学习的方法)和重复法等。在基于光流算法的方法中,Liao 等[72]用各种光流方法产生高分辨率候选对象,并被卷积神经网络集合起来。VSRnet[73]和 CVSRnet[74]通过 Druleas 算法[75]实施运动补偿,用卷积神经网络取连续帧作为输入,并预测高分辨率帧。Liu 等[76-77]完成了光流校正,并提出了一个时间自适应网络,产生各种时间尺度的高分辨率帧,并自适应聚合。此外,其他人也试着直接学习运动补偿。VESPCN[78]利用一个可训练的空间转换器[79]学习基于临近帧的运动补偿,并使多帧进入一个用于端对端预测的空时 ESPCN。此外,Tao 等[80]植根于精确的低分辨率成像模型,提出一个亚像素类模块,同时实现运动补偿和超分辨率,这样更有效地融合对齐的帧。

另一个趋势是使用重复法获得空-时约束,无外在运动补偿。特别地,BRCN[81]采用一个双向架构,并用卷积神经网络、重复神经网络和有条件的卷积神经网络,分别模拟空间、时间和空-时依赖。类似地,STCN[82]用一个深度卷积神经网络和一个双向 LSTM[83]提取空间和时间信息。此外,FRVSR[84]用以前推断的高分辨率估计,通过两个深层卷积神经网络以重复方式重构后续高分辨率帧。

除了上述文献,FAST[84-85]利用紧凑的压缩算法描述提取结构和像素相关,把一帧的超分辨率转到临近帧,使最先进的超分辨率算法加速15倍,而性能损失则很小(0.2dB)。

此外,Jo等[86]产生动态上采样滤波器和基于每个像素的局部空时邻近高分辨率残留图像,同时避免外在运动补偿。

9.4.5 高效度量及其他评估策略

评估度量是机器学习最基础的分量之一。如果不能精确地测量模型性能,研究人员验证改进将有很大的难度[87]。超分辨率度量面对如此的挑战,需要更多探索。

(1)更精确的度量:使用最广泛的超分辨率度量法是 PSNR 和 SSIM。然而,PSNR 趋于导致过度的平滑,结果在几乎不可辨别的图像间变化常常很大。根据亮度、对比度和结构,SSIM 虽能评估图像,但是仍不能精确地测量视觉图像质量。此外,平均意见分数(MOS)最接近人的视觉反应,但是需要许多努力,且不可再生。这样,急需更精确的重构质量评估度量。

(2)盲 IQA 法:目前,超分辨率所用的多数度量法是全引用法,即假设低分辨率-高分辨率图像均具有优良的质量。

然而,因为难以得到此类数据集,所以评估常用数据集通过手动退化实施。在此情况下,评估任务是预定退化的逆过程。因此,开发盲 IQA 法也有很大的需求。最近,基于卷积神经网络的方法取得了图像超分辨率的巨大成功。然而,多数基于深层卷积神经网络的超分辨率模型尝试提高失真测量(如 PSNR、SSIM、IFC、VIF),同时带来不良的量化知觉质量(如人情评分、无参考质量法(如 NIQE))。少数文献尝试以降低失真测量性能为代价提高知觉质量。最新研究显示,失真和知觉质量两者之间彼此不统一,但两者之间总有平衡。在知觉质量方面出众的复原算法,在失真测量方面则常常很差。本章分析用于单帧图像超分辨率问题的失真和知觉质量间的平衡,称为增强深层超分辨率(EDSR)网络,使其适合一个特殊的失真测量范围,以实现更好的知觉质量。尽管训练增强深层超分辨率的原始网络以使基于每个像素精度定义的误差最小,但是网络用一个生成的对手网络(GAN)架构[88-89]训练,增强深层超分辨率作为生成模块[90]。参考文献[87]中的工作称为增强知觉超分辨率(EPSR)网络,综合 MSE 损失、知觉损失和对手损失对其训练。试验表明,增强知觉超分辨率实现了失真和知觉质量间的平衡,而现有的方法在这些测量的任何方面都完成得很好。

9.4.6 多重先验

模糊通常通过一些已知分布模拟,但是与实际情况相去甚远。理想地,模糊类型需与另一模型参数一起估计。模糊识别可大大提高超分辨率重构,并可同时图像复原。

训练阶段,混合先验模型可通过把与低分辨率图像关联的大的非线性特征空间转换到一系列线性子空间来简化。首先,用一个新颖的斑块识别处理机制(依赖低分辨率斑块不同的弯曲将图像划分成图像斑块。其次,学习对应每一组的混合先验模型。此外,因为不同的先验分布在超分辨率中可能起各种不同的作用,因此可以选择一些分布并组合,具有比通常的高斯先验更强大的性能。在测试阶段,学习的混合先验模型帮助把起作用的输入低分辨率特征映射到适当的子空间。最后,以一个新的混合相当方法重构分辨率相当高的画面。试验结果表明,该方法是定量又定性的,比一些同代的超分辨率方法更好[91]。

9.4.7 正则化

由于观测的低分辨率帧不足,极易引起超分辨率问题。解决该问题的一个经典方法是使用不同的先验,被称为正则化[92]。

一个经典或常见的先验是基于 Tikhonov 模型[9,46,92-94]。该先验引入复原问题去除图像外来噪声的平滑度约束。Tikhonov 先验法的缺点是它倾向于破坏边缘——一种图像退化效应。因此,先验已激发许多研究人员的兴趣,开发了同时抑制噪声并保持关键图像特征的模型:Huber Markov 随机场(Huber – RMF)[95-96]、边缘自适应 RMF[97-98]、粗略方向[99-100] 和总变化(TV)[5,75,93,101-103]。

上述先验方法中,总变化吸引了研究人员更多的关注,因为产生令人满意的目标和主观品质结果。总变化模型的主要缺点是阻塞和阶梯效应,在零点虚边缘生成接近边缘和不可微性,这是使数字实现具有相当挑战性的一种情形。总变化模型最初被用于图像去噪[104]。随后,该模型被用于其他应用:超分辨率、MRI 医学图像重构[105]、修补[106-107] 和去模糊[108]。本章研究了一些解决超分辨率问题的基于总变化的经典方法[53,101,105]。

2008 年,Marquina 和 Osher 基于受约束的总变化架构[101],提出了一个解决超分辨率问题的卷积模型。在其著作中,作者介绍了作为提高空间分辨率迭代求精步骤的 Bregman 算法[109-114]。Bregman 发散的使用允许更包罗万象的模型[115]。结果演示了产生细节和更清晰图像的方法,但是阻塞和阶梯假象仍很明显。在参考文献[103]中,对于多帧超分辨率(MFSR)问题[116-118],Farsiu 及其同事提出了一个双向总变化(BTV)先验,基于 L1 限额(标准)最小化和双向滤波器正则化函数。他们的方法计算合理,而且对运动和模糊估计所致的误差稳健,并令人信服地产生更清晰的图像。此外,该方法的双向滤波器还引入假象,如阶梯和梯度颠倒。此外,双向总变化并未完美地处理局部图像平滑度[119]。同时,通过数字方法实现 L1 标准很麻烦,因为它们可隐藏超分辨率数据项。总变化先验解决了高分辨率图像重构中的下列问题:噪声、模糊、区域缺失、压缩假象和运动估计误差[70-71,120]及几种运动类型和场景退化中的作用,试验结果战胜了其他经典的超分辨率方法。

Ren 等提出了一个超分辨率方法,该方法基于分数阶总变化正则化,重点是处理精细的细节,如图像中的纹理[121]。结果显示,某种程度上他们的方法克服了传统总变化的缺点。

在参考文献[21]中,Li 等尝试通过提出两种解决全局总变化缺点的方法,即局部自适应总变化和梯度一致性,保证边缘更清晰、平坦区更平滑。该方法严重依赖图像的梯度变化细节,这是一个在噪点均匀区可能产生伪边缘的特征。注意,噪声和边缘是具有高梯度(或高亮度)值的图像特征。因为 Li 的方法是依赖梯度的,所以它同等对待噪声和边缘,这或产生不必要的假象。

Yuan 等提出了一个用于多帧超分辨率问题的空间加权的总变化模型[122],他们的模型合并了局部识别个别像素空间特性的空间信息指示器(差值曲率)。这样,具备必要的正则化水平。Yuan 等利用 the majori sation – 最小化算法优化公式,结果表明,Yuan 等的方法克服了原始总变化模型的一些挑战(阻碍分段常数解),且不太灵敏。

9.4.8 新颖体系架构

随着卷积神经网络的发展,单帧图像超分辨率已取得了空前突破[123-124]。这些方法大部分增加网络的深度以得到更大的接收场。然而,本章发现:盲目地堆叠特征图及简单的层叠式结构,对于单帧图像超分辨率,完全无瑕的卷积神经网络并不能在超分辨率重构中实现高的利用率。虽然假设基于不同深度或相同深度的特征图,但是在图像重构过程中不同通道具有不同的作用 – 压缩和激励网络评估不同特征图的重要性,同时构建网络[125]。此外,也在架构中进行密集连接操作,以更好地使用前后关系信息和特征图。广泛的试验验证了提出的方法可提高复原性能,并在超分辨率任务中实现了最先进的结果。

稀疏编码已广泛应用于基于学习的单帧图像超分辨率,并通过共同学习有效表示低分辨率和高分辨率图像斑块对,得到有希望的性能。然而,由于强大而胜任低分辨率图像斑块表示的 Ad – hoc 假设是线性的,产生的高分辨率图像常常出现振铃、参差不齐和模糊假象,取决于多种类似于或具有与对应高分辨率图像斑块表示相同的支撑集。受深度学习成功的激励,提出了一个称为耦合深层自动编码器(CDA)的单帧图像超分辨率数据驱动模型[126]。耦合深层自动编码器基于一个新的深层架构,并有高的表示能力。耦合深层自动编码器同时学习低分辨率和高分辨率图像斑块内在表示和大数据驱动函数(精确地把这些低分辨率表示映射到他们对应的高分辨率表示)。广泛的试验验证了与其他先进的 Set5 和 Set14 数据集方法相比,耦合深层自动编码器对单帧图像超分辨率卓越的效能和功效。

参考文献[127]解决了对基于 sketch 的图像恢复(SBIR)学习稳健的跨域表示问题。尽管多数 SBIR 法的重点是提取用于直接特征匹配的低级和中级描述符,然而近来的文献表明,学习耦合特征表示的好处是描述来自两个相关源的数据。然而,典型地跨域表示深度学习法转化成难以优化、导致不令人满意性能的非表面弯曲最小化问题。灵感来自自定进度学习(一种旨在克服与局部优化相关的收敛问题学习法)。通过意味深长的顺序(从易到难)利用采样,提出了 cross-paced 部分课程学习(CPPCL)架构。与现有仅考虑一个模态且不能处理先验知识的自定进度学习法相比,CPPCL 特别设计评估学习进度,通过联合处理双源数据和以单独课程形式提供特定模态先验信息。另外,得益于学习词典,CPPCL 嵌入稳健的 SBIR 耦合表示,该方法是广泛评估四个公开可用的数据集(表现出比 SBIR 法更卓越的性能)。

OctNet 是一个以稀疏三维数据的深度学习表示[128]。与现有模型相比,这种表示赋能三维卷积神经网络,它既深又具有高分辨率。为此目标,完成把稀少数据输入到用一组不均衡八权树分级划分的空间。这里每个叶节点存储一个合并的特征表示。允许把内存分配和计算重点放到相应的密集区,并赋能更深层网络,无分辨率的妥协。OctNet 表示法的效用通过分析分辨率对几项三维任务(包括三维目标分类、方位估计和点云标注)的影响来体现[129]。

课程学习[130]提到从较容易的小任务开始,逐渐加大任务难度。因为超分辨率本质上是一个不适定问题和一些恶劣的条件,如大标定因数,噪声或模糊进一步增加了学习难度。因此,课程训练策略对该问题助益良多。考虑到一步完成大的超分辨率是一项非常困难的任务,Wang 等[47]和 Bei 等[131]分别提出了 ProSR 和 ADRSR,它们是进步的,不仅在体系架构方面(3.1.3 节),而且在训练程序上。训练从两个上采样开始,完成当前部分训练后,逐渐设置四个部分或较大的标定因数,并与以前的部分相融合。特别是,ProSR 融合了两部分,通过线性融合该级输出和以前级别的上采样输出[132],ADRSR 则连接它们并附加另一卷积层。相反,Park 等[71]把 8 倍超分辨率问题分成三个小问题(1~2 倍超分辨率,2~4 倍超分辨率,4~8 倍超分辨率),并对每个问题进行单独网络训练。其中两个被分解,并共同调整,然后对另一个。此外,在困难条件下它们还把 4 倍超分辨率分解成三个小问题(去噪/去模糊,1~2 倍超分辨率,2~4 倍超分辨率),并采用类似的训练策略。与常见的训练程序相比,该课程学习策略不仅大大降低训练难度和以所有标定因数提高性能(特别是大的因数),而且大大缩短总的训练时间。

尽管用更快更深的卷积神经网络实现了单帧图像超分辨率的精度和速度突破,然而一个悬而未决的中心问题是:如何用大标定因数超分辨率恢复更精细的纹理细节?目标函数选择主要推动基于最优超分辨率程序的行为。最近的工作重点放在使均方重构误差最小化。评估虽具有高的峰值信噪比,但是常常因缺少高频细节,知觉上并不那么令人满意。从某种意义上说,它们未能匹配更高的分辨率保真度期待。SRGAN 是一个能够推断 4 倍标定因数真实感自然图像的超分辨率的 GAN。它采用一个知觉损失函数,由对手的损失和内容损失组成。对手的损失推动我们用训练区分高分辨率图像和原始真实感图像的鉴频网络解决多种自然图像。另外,它采用被知觉相似性而不是像素空间相似性推动的内容损失。深度残留网络能够在公共基准上从严重下采样图像恢复真实感纹理。使用 SRGAN 进行的广泛 MOS 试验表明,在知觉质量中有非常大的增益。用 SRGAN 得到的 MOS 评分比用同时代的任何方法得到的结果都更加接近原始高分辨率图像[133]。

深层卷积神经网络已得到广泛使用,并已在许多图像或视频处理和分析任务中实现最先进的性能。尤其是,对于图像超分辨率处理,当与浮浅的基于学习的方法相比时,以前基于卷积神经网络的方法已带来重大改进。然而,当应用于遥感卫星图像超分辨率时,以前基于卷积神经网络的算法简单直接或跳跃连接性能不佳。在该研究中,一个简单而有效的卷积神经网络架构,即深层蒸馏递归网络(DDRN)[134],被用于显示视频卫星图像超分辨率。DDRN 包括一组超密残留块(UDB),一个多尺度净化单元(MSPU)和一个重构模块。尤其是,通过在每个 UDB 多通道单元内和它们之间增加充足的交互式链路,可有效地共享多个平行卷积层提取的特征。与经典的基于密度连接的模型相比,DDRN 具有以下主要特性。

(1) DDRN 包含有相同卷积层的更多连接节点。

(2) DDRN 构建了一个在网络不同阶段完成特征蒸馏和补偿的蒸馏和补偿机制。尤其是,信息传播过程中丢失的高频部分可在 MSPU 中得到恢复。

(3) 最终高分辨率图像可受益于 UDB 提取的特征图,并补偿从 MSPU 得到的。Kaggle 开源数据集和 Jilin – 1 视频卫星图像试验说明 DDRN 胜过传统的基于卷积神经网络的基线和一些最先进的特征提取方法。

9.4.9 3D 超分辨率

9.4.9.1 深度图超分辨率

深度图记录浏览场景中点和物体间的距离,在许多任务中深度信息起很重要的作用,例如:姿态估计[89,123-124]、语义分割[135-136]等。然而,由于生产性和经济限制,深度传感器产生的深度图常常是低分辨率的,并经历恶化效应,如噪声、量化、丢失值等。这样,引入超分辨率以提高深度图的空间分辨率。今天,一种最流行的深度图超分辨率实践,是用另一个经济的 RGB 相机得到相同场景的高分辨率图像,用于引导超分辨低分辨率深度图。特别是,Song 等[137]利用深度场统计和深度图与 RGB 图像之间的局部相关约束全局统计和局部结构。Hui 等[138]首先利用两个卷积神经网络上采样的深度图和下采样的高分辨率 RGB 图像;然后 RGB 特征用作相同分辨率上采样过程指导。类似地,Ni 等[139]和 Zhou 等[114]用高分辨率 RGB 图像作为引导,分别提取高分辨率边缘图,并预测失去的高频元件。Xiao 等[140]用锥形网络扩大接收场,分别从低分辨率深度图和高分辨率 RGB 图像中提取特征,并融合这些特征预测高分辨率深度图。Haefner 等[141]通过采用明暗恢复图像技术,完全利用色彩信息引导超分辨率。与上述文献相比,Riegler 等[142]把卷积神经网络与一个能量最小化模型相结合、以一种强大变化模型的形式恢复高分辨率深度图,无其他基准图像。

在触发模式,用一台手持式相机从一组不均匀的运动-模糊低分辨率图像获得潜在三维场景的高分辨率图像估计问题出现在[i1,i2,h5]中。现有的解决运动模糊的盲超分辨率技术限于前平行平面场景。最初,设想了一个超分辨率运动模糊模型解释三维场景中图像形成的过程。随后,该模型已用来确定三个未知项:相机轨迹、深度图和潜在高分辨率图像。最后,从斑块恢复对应每个低分辨率观测全局高分辨率相机的运动,取决于输入图像中的基准深度层。用估计的轨迹,反复使用一个交互的最小化架构得到潜在的高分辨率图像和深度图。综合真实数据试验表明,提出的方法远远优于最先进的技术。

单一目标选择性-平面照射显微镜(soSPIM)由与安装在一个标准反转显微镜上的激光束控单元组合的微镜腔实现,照射和探测通过相同的目标进行。soSPIM 可与标准采样准备、特征高背景抑制及高效光子采集一起使用,允许基于整个单元或单元集合的三维单分子的超分辨率成像,较大的反射镜能够扩大系统能力[143]。

由于荧光取样器的高背景噪声和快速光褪色,用超分辨率荧光显微镜(SR-FM)的厚采样三维重构仍具有挑战性[144],超分辨率荧光显微镜可重构具有高局部化精度和无光褪色问题的厚采样三维结构。背景噪声用线扫共焦显微镜通过光学部分采样降低,光褪色问题正用 DNA-PAINT 克服(纳米尺度高度图中成像点累积)。

9.4.10 深度学习和计算智能

虽然深层网络已表现出异常的超分辨率任务性能,但是仍存在几个开放的

第9章 深度学习作为无人机系统中超分辨率成像的可替代方案

研究问题。我们将这些未来的研究方向罗列如下。

(1)结合先验知识:当前的超分辨率深层网络是以端对端学习的数据驱动模型。虽然该方法总体上展现出优良的结果,但是,当出现特别等级的退化时,因为不存在大量训练数据(如医学成像中),证明其性能欠佳。在此情况下,如果知道传感器、成像物体/场景和捕获条件信息,可设计有用的先验得到高分辨率图像。为了实现更好的超分辨率,重点关注该方向提出基于深层网络[130]和粗编码[131]先验的新文献。

(2)目标函数和度量:现有的超分辨率法突出使用像素级误差测量法(如 l_1、l_2 距离),这是另一标准或综合度量方法。因为这些措施仅压缩局部像素级信息,因此产生的图像并不总是知觉地提供声音结果。例如,表明高峰值信噪比和SSIM 值图像给出过度平滑的低知觉质量图像[132]。为克服这一问题,文献提出了几种解决知觉损失的措施。传统的知觉度量是很确定的,如 SSIM[145]、多尺度SSIM[146];最新的方法是学习模拟人类感知图像,如 LPIPS[147]和 PieAPP[148]。这些措施中每一个都有失误的情况。结果,并没有在各种条件下都工作最佳并完美量化图像质量的通用知觉度量标准。因此,新目标函数的发展是一个开放的研究问题。

(3)需要统一的方案:现实中,常常同时出现两种或多种退化。在这些情况下,重要的考虑是如何共同恢复分辨率更高、噪声低和更多细节的图像。目前,为超分辨率开发的模型通常仅限于在其他退化面前会遇到的一种情况。此外,特定问题模型的体系架构、损失函数和训练细节不同。设计同时完成好几个低级视觉任务的统一模型是一项挑战[124-135]。

(4)无监督学习的图像超分辨率:该研究中讨论的模型通常考虑低分辨率-高分辨率图像对学习超分辨率的映射函数。一个有趣的方向是研究在相应高分辨率图像不可用的情况下如何完成超分辨率。该问题的一个解决方案是零样本超分辨率(学习一个给定图像的进一步下采样超分辨率模型)。然而,当输入图像分辨率较低时,该方案就无法奏效。无人管理的图像超分辨率计划通过学习不成对的低分辨率-高分辨率图像组的函数来解决该问题。这样的能力对于实际寿命设定非常有用,因为在几种情况下得到匹配的高分辨率图像并非微不足道[149]。

(5)更高的超分辨率速率:当前的超分辨率模型通常不能应付极端的超分辨率(可用于以下情况,如在拥挤场景中提高人脸识别率),极少文献的目标是超分辨率高于 8 倍(如 16 倍和 32 倍)。在极端上采样条件下,保持图像中精确的局部细节变得具有挑战性。此外,一个开放的问题是如何保持超分辨图像的高知觉质量。

(6)任意超分辨率:对于给定的输入,不切实际的想定常常不知道哪个上采样因子最佳。当数据集中所有图像下采样因数未知时,这就成为训练中的一个重大挑战,因为一个模型难以实现几级细节压缩。在这些情况下,重要的是在训练前首先描述退化水平,并通过指定的超分辨率模型完成推理。

(7)真实与假退化:现有超分辨率著作通常用双三次插值产生低分辨率图像。与用双三次插值合成产生的图像相比,现实想定中遇到的真实低分辨率图像有不同的分布。结果表明,在实际想定中训练人造退化的超分辨率网络对真实低分辨率图像归纳不充分,解决该问题的一个最新的努力是首先学习一个真实退化的 GAN 模型[149]。

9.4.11 网络设计

适当的网络设计不仅确定一个具有卓越性能最大值的假设空间,而且支持高效的数据学习及数据表示,没有过分的空间和计算冗余。下面讨论一些有前景的网络改进方向。

(1)融合局部和全局信息:大的接收场提供更多有助于产生更逼真高分辨率图像的前后关系信息。有希望融合局部和全局信息,提供超分辨率不同比例的背景信息[150-155]。

(2)融合低级和高级信息:深层卷积神经网络中的浅层倾向于提取低级特征,如色彩和边缘,而更深层提取更高级的表示,像物体特性。这样,融合低级细节与高级摘要语义,可极大地帮助高分辨率重构。

(3)特定的环境关注:不同的环境聚焦不同的超分辨率信息。例如,草坪区可能更多关心色彩和纹理,而动物身体区可能更多聚焦毛发细节。因此,合并关注机制以利用前后关系信息提高对关键特征的关注,可促进逼真细节的产生。

(4)轻量化体系结构:现有的超分辨率模式倾向于追求终极性能,同时忽略模型尺寸和推理速度。例如,在 TitanGTX[114] 上,DIV2K[137] 每幅图像的 4 倍超分辨率,EDSR[136] 耗时 20s,DBPN[138] 处理 8 倍超分辨率需要 35s[139]。实际应用中,如此长的预测时间是不可接受的。这样,轻型体系结构是必要的。如何减小模型尺寸和加速预测,同时又保持性能仍是一个问题。

(5)上采样层:尽管上采样运算在超分辨率中起非常重要的作用,但是现有的方法或多或少都有缺点:基于内插的方法带来代价很大的计算问题,且不能端对端学习,变换卷积产生棋盘格假象,亚像素层带来不均匀的接收场分布。因此,仍需研究如何完成有效而高效的上采样,特别是高标定因数。

然而,消耗性地使用超分辨率仍是个问题。一个特殊的瓶颈出现在多帧超分辨率架构,致使研究人员紧张地寻找基于单幅图像实例的超分辨率。因此,单图像超分辨率算法的性能以外部数据库的可信度为转移,仍需自适应更强、更先进和具有深远适应性的速度更快的技术。

硬件的快速进步也将挑战超分辨率架构的新应用。例如,谷歌 Skybox 计划将用遥感超分辨率成像实时传送高分辨率。超分辨率还可扩展 FM 和多基线断层合成孔径雷达成像。但是,由于计算性能欠佳和时间消耗,以及大规模应用需要的必要加速度策略,在 UAV – CPS 中超分辨率方法的可行性仍受到限制。

9.5 小结

本章的目的是获得无人机超分辨的图像。本章提出了一个基于卷积神经网络的超分辨率架构,用于建立在十分广泛的目录调查基础之上的低分辨率图像[156-164]。从分析可以看到,提出的基于卷积神经网络的架构可有效地超分辨无人机航拍图像,在 MSE、PSNR、SSIM 和 VIFP 方面,与先进的算法相比,具有更好的性能。

超分辨率成像可避开或补偿成像架构内在的硬件和通信限制,提供一幅带有更好的有帮助内容的清晰图像[165-188]。超分辨率还可在前端预处理阶段加速扩展各种计算机视觉应用中对性能方面提供帮助。

其他神经元启发式方法也可在深层体系架构中帮助预处理数据,从而提高深度学习架构速度和减小计算负荷[189-192]。

参考文献

[1] S. K. Seelan, S. Laguette, G. M. Casady, and G. A. Seielstad, Remote sen – sing applications for precision agriculture: a learning community approach, Remote Sensing and Environment, 88(1), 157 – 169, 2003.

[2] J. Everaerts, The use of unmanned aerial vehicles (UAVs) for remote sensing and mapping, The International Archives of the Photogrammetry, Remote Sensing and Spatial Information Sciences, 37(2), 1187 – 4, 2008.

[3] G. Grenzdörffer, A. Engel, and B. Teichert, The photogrammetric potential of low – cost UAVs in forestry and agriculture, The International Archives of the Photogrammetry, Remote Sensing and Spatial Information Sciences, 31(3), 1207 – 1213, 2008.

[4] V. V. Estrela, O. Saotome, H. J. Loschi, *et al.*, Emergency response cyber – physical framework for landslide avoidance with sustainable electronics, Technologies, 6, 42, 2018. doi: 10.3390/technologies6020042.

[5] M. A. de Jesus, V. V. Estrela, and O. Saotome, Super – resolution in a nutshell, Proceedings BTSym 2016, Campinas, SP, Brazil.

[6] Y. Tang, P. Yan, Y. Yuan, and X. Li, Single – image super – resolution via local learning, Inter-

national Journal of Machine Learning and Cybernetics,2,15 – 23,2011.

[7] W. Wu,Z. Liu,X. He,and W. Gueaieb,Single – image super – resolution based on Markov random field and contourlet transform,Journal of Electronic Imaging,20(2),023005 – 023005,2011.

[8] K. I. Kim,and Y. Kwon,Single – image super – resolution using sparse regression and natural image prior,IEEE Transactions on Pattern Analysis and Machine Intelligence,32(6),1127 – 1133,2010.

[9] A. M. Coelho,and V. V. Estrela,A study on the effect of regularization matrices in motion estimation,International Journal of Computer Applications,51(19),17 – 24,2012.

[10] W. T. Freeman,T. R. Jones,and E. C. Pasztor,Example – based super – resolution,IEEE Computer Graphics and Applications,22(2),56 – 65,2002.

[11] K. Zhang,X. Gao,X. Li,and D. Tao,Partially supervised neighbor embedding for example – based image super – resolution,IEEE Journal of Selected Topics in Signal Processing,5(2),230 – 239,2011.

[12] C. Kim,K. Choi,and J. B. Ra,Example – based super – resolution via structure analysis of patches,IEEE Signal Processing Letters,20(4),407 – 410,2013.

[13] Z. Xiong,D. Xu,X. Sun,and F. Wu,Example – based super – resolution with soft information and decision,IEEE Transactions on Multimedia,15(6),1458 – 1465,2013.

[14] C. Y. Yang,J. B. Huang,and M. H. Yang,Exploiting self – similarities for single frame super – resolution,in Proc Computer Vision – ACCV 2010.(Springer Queenstown,New Zealand,2011),pp. 497 – 510.

[15] O. Mac Aodha,N. D. Campbell,A. Nair,and G. J. Brostow,Patch based synthesis for single depth image superresolution,in Proc. Computer Vision – ECCV(Springer Florence,Italy,2012),pp. 71 – 84.

[16] X. Gao,K. Zhang,D. Tao,and X. Li,Joint learning for single – image super – resolution via a coupled constraint,IEEE Transactions on Image Processing,21(2),469 – 480,2012.

[17] J. Li,and X. Peng,Single – frame image super – resolution through gradient learning,in Proc 2012 International Conference on Information Science and Technology(ICIST)(IEEE Hubei,2012),Wuhan,China pp. 810 – 815.

[18] K. Zhang,X. Gao,D. Tao,and X. Li,Single image super – resolution with non – local means and steering kernel regression,IEEE Transactions on Image Processing,21(11),4544 – 4556,2012.

[19] M. Yang,and Y. Wang,A self – learning approach to single image super – resolution,IEEE Transactions on Multimedia,15,498 – 508,2013.

[20] K. Zhang,X. Gao,D. Tao,and X. Li,Single image super – resolution with multiscale similarity learning,IEEE Transactions on Neural Networks and Learning Systems,24(10),1648 – 1659,2013.

[21] X. Li,Y. Hu,X. Gao,D. Tao,and B. Ning,A multi – frame image super – resolution method,Signal Processing,90(2),405 – 414,2010.

[22] B. J. Maiseli,Q. Liu,O. A. Elisha,and H. Gao,Adaptive Charbonnier superresolution method

with robust edge preservation capabilities, Journal of Electronic Imaging, 22(4), 043027, 2013.

[23] B. Maiseli, O. Elisha, J. Mei, and H. Gao, Edge preservation image enlargement and enhancement method based on the adaptive Perona – Malik nonlinear diffusion model, IET Image Processing, 8, 753 – 760, 2014.

[24] X. D. Zhao, Z. F. Zhou, J. Z. Cao, et al. , Multi – frame super – resolution reconstruction algorithm based on diffusion tensor regularization term, Applied Mechanics and Materials, 543, 2828 – 2832, 2014.

[25] J. Yang, and T. Huang, Image super resolution: historical overview and future challenges. In Super – Resolution Imaging, pp. 1 – 24. CRC Press, Boca Raton, FL, USA, www. ifp. illinois. edu/ ~ jyang29/papers/chap1. pdf – United States.

[26] J. Sun, J. Sun, Z. Xu, and H. Y. Shum, Image Super – Resolution Using Gradient Profile Prior, IEEE Conference on Computer Vision and Pattern Recognition, Anchorage, AK, USA, 2008.

[27] W. S. Tam, C. W. Kok, and W. C. Siu, Modified edge – directed interpolation for images, Journal of Electronic Imaging, 19(1), 1 – 20, 2010.

[28] L. Wang, S. Xiang, G. Meng, H. Wu, and C. Pan, Edge – directed singleimage super – resolution via adaptive gradient magnitude self interpolation, IEEE Transactions on Circuits and Systems for Video Technology, 23, 8, 1289 – 1299, 2013.

[29] M. Bertero, and P. Boccacci. Introduction to Inverse Problems in Imaging, CRC Press, Boca Raton, Florida, USA, 1998.

[30] M. V. Joshi, S. Chaudhuri, and R. Panuganti, A learning – based method for image super – resolution from zoomed observations, IEEE Transactions on Systems, Man, and Cybernetics—Part B: Cybernetics, 35(3), 527 – 537, 2005.

[31] M. V. Joshi, S. Chaudhuri, and P. Rajkiran, Super – resolution imaging: use of zoom as a cue, Image and Vision Computing, 14(22), 1185 – 14, 2004.

[32] K. Zhang, X. Gao, D. Tao, and X. Li, Single image super – resolution with nonlocal means and steering kernel regression, IEEE Transaction on Image Processing, 21, 4544 – 4556, 2012.

[33] X. Gao, Q. Wang, X. Li, D. Tao, and K. Zhang, Zernike – moment – based image super – resolution, IEEE Transaction on Image Processing, 20, 2738 – 2747, 2011.

[34] M. Irani, and S. Peleg, Improving resolution by image registration, CVGIP Graphical Models and Image Processing, 53, 231 – 237, 1991.

[35] C. Dong, and C. C. Loy, Image super – resolution using deep convolutional networks, IEEE Transactions on Pattern Analysis and Machine Intelligence, 38(2), 295 – 307, 2016.

[36] J. Chen, and X. He, Single image super – resolution based on deep learning and gradient transformation, IEEE 13th International Conference on Signal Processing (ICSP), Chengdu, China, 2016.

[37] Y. Yoon, and H. G. Jeon, Learning a deep convolutional network for lightfield image super – resolution, IEEE International Conference on Computer Vision Workshop, Santiago, Chile, 2015.

[38] Z. Wang, and D. Liu, Deep networks for image super – resolution with sparse prior, IEEE International Conference on Computer Vision (ICCV), Santiago, Chile, 2015.

[39] Y. Kato, and S. Ohtani, Image super-resolution with multi-channel convolutional neural networks, IEEE International New Circuits and Systems Conference (NEWCAS), Vancouver, BC, Canada, 2016.

[40] A. Deshpande, and P. Patavardhan, Multiframe super-resolution for long range captured iris polar image, IET Biometrics, 6(2), 108, 2017.

[41] M. Haghighat, H. Seyedarabi, and A. Aghagolzadeh, Multi-focus image fusion for visual sensor networks in DCT domain, Computers and Electrical Engineering, 37(5), 789-797, 2011.

[42] R. Fattal, Single Image Dehazing, Proc. ACM SIGGRAPH'08, Los Angeles, CA, USA, 2008.

[43] R. Tan, Visibility in Bad Weather from a Single Image, Proc. IEEE Conf. Computer Vision and Pattern Recognition, Anchorage, Alaska, USA, 2008.

[44] J.-B. Wang, H. Ning, Z. Lu-Lu, and L. Ke, Single image dehazing with a physical model and dark channel prior, Neurocomputing, 149, 718-728, 2015.

[45] Z. Li, Zhengguo, and J. Zheng, Edge-preservingdecomposition-basedsingle image haze removal, IEEE Transactions on Image Processing, 24, 5432-5441, 2015.

[46] N. Nguyen, P. Milanfar, and G. Golub, A computationally efficient super-resolution image reconstruction algorithm, IEEE Transactions on Image Processing, 10(4), 573-583, 2001.

[47] Z. Wang, A. Bovik, H. Sheikh, and E. Simoncelli, Image quality assessment: from error visibility to structural similarity, IEEE Transactions on Image Processing, 13(4), 600-612, 2004.

[48] S. Bovik, and G. Veciana, An information fidelity criterion for image quality assessment using natural scene statistics, IEEE Transaction on Image Processing, 14(4), 2117-2128, 2014.

[49] T. Lukes, K. Fliegel, and M. Klima, Performance Evaluation of Image Quality Metrics with Respect to their Use for super-resolution Enhancement, Fifth International Workshop on Qualcomm Multimedia Experience, 2013.

[50] X. Zhou, and B. Bhanu, Evaluating the Quality of Super-Resolved Images for Face Recognition, IEEE Computer Society Conference on Computer Vision and Pattern Recognition Workshops, 2008.

[51] Z. Wang, and A. C. Bovik, Mean squared error: love it or leave it? A new look at signal fidelity measures, IEEE Signal Processing Magazine, 26, 98-117, 2009.

[52] A. Tanchenko, Visual-PSNR measure of image quality, Journal of Visual Communication and Image Representation, 25(5), 874-878, 2014.

[53] A. M. Coelho, V. V. Estrela, F. P. Carmo, and S. R. Fernandes, Error Concealment by Means of Motion Refinement and Regularized Bregman Divergence, Proceedings of the 13th International Conference on Intelligent Data Engineering and Automated Learning, Natal, Brazil, 2012.

[54] A Punnappurath, T. M. Nimisha, and A. N. Rajagopalan, Multi-image blind super-resolution of 3D scenes, IEEE Transactions on Image Processing, 26, 5337-5352, 2017.

[55] S. Takemura, Optimize deep super-resolution and denoising for compressed textures. SIGGRAPH Asia Posters, Tokyo, Japan, 2018.

[56] G. Kim, and N. Baek, A Height-Map Based Terrain Rendering with Tessellation Hardware. 2014 International Conference on IT Convergence and Security (ICITCS), Beijing, China, 2014, pp. 1-4.

第 9 章　深度学习作为无人机系统中超分辨率成像的可替代方案

[57] S. R. Fernandes, V. V. Estrela, Hermes A. Magalhaes, and O. Saotome, On Improving Sub-Pixel Accuracy by Means of B-Spline, Proceedings of the 2014 IEEE International Conference on Imaging Systems and Techniques (IST 2014), Santorini, Greece, pp. 68-72, ISBN: 9781479952199.

[58] F. D. Silva, V. V. Estrela, and L. J. Matos, Hyperspectral Analysis of Remotely Sensed Images, Ed.: Carolina Biblio, Oliver Hensel, Jeferson Francisco Selbach (Org.). Sustainable Water Management in the Tropics and Subtropics-and Case Studies in Brazil. 1ed., Kassel, Germany: University of Kassel, 2011, vol. 2, pp. 398-423, ISBN: 9788563337214.

[59] M. Fauvel, Y. Tarabalka, J. A. Benediktsson, J. Chanussot, and J. C. Tilton, Advances in the spectral-spatial classification of hyperspectral images, Proceedings of the IEEE, 101, 652-675, 2013.

[60] Y. Fu, Y. Zheng, I. Sato, and Y. Sato, Exploiting spectral-spatial correlation for coded hyperspectral image restoration, in Proc. 2016 Conference on Computer Vision and Pattern Recognition (CVPR 2016), Las Vegas, NV, USA, 2016.

[61] B. Uzkent, M. J. Hoffman, and A. Vodacek, Real-time vehicle tracking in aerial video using hyperspectral features, in Proc. 2016 IEEE Conference on Computer Vision and Pattern Recognition Workshops (CVPRW2016), Las Vegas, NV, USA, 2016.

[62] B. Uzkent, A. Rangnekar, and M. J. Hoffman, Aerial vehicle tracking by adaptive fusion of hyperspectral likelihood maps, in Proc. 2017 IEEE Conference on Computer Vision and Pattern Recognition Workshops (CVPRW2017), Honolulu, HI, USA.

[63] W. Huang, L. Xiao, Z. Wei, H. Liu, and S. Tang, A new pansharpening method with deep neural networks, in IEEE Geoscience and Remote Sensing Letters, vol. 12, 5, 1037-1041, 2015.

[64] G. Masi, D. Cozzolino, L. Verdoliva, and G. Scarpa, Pansharpening by convolutional neural networks, Remote Sensing, 8, 594, 2016.

[65] C. Dong, C. C. Loy, K. He, and X. Tang, Learning a deep convolutional network for image super-resolution, in Proc. 13th European Conference on Computer Vision (ECCV 2014), Zurich, Switzerland, 2014.

[66] Y. Wei, Q. Yuan, H. Shen, and L. Zhang, Boosting the accuracy of multi-spectral image pansharpening by learning a deep residual network, IEEE Geoscience and Remote Sensing Letters, 14, 10, 1795-1799, 2017.

[67] K. He, X. Zhang, S. Ren, and J. Sun, Deep residual learning for image recognition, in Proc. 2016 Conference on Computer Vision and Pattern Recognition (CVPR 2016), Las Vegas, NV, USA, 2016.

[68] Y. Qu, H. Qi, and C. Kwan, Unsupervised sparse Dirichlet-net for hyperspectral image super-resolution, in Proc. 2018 IEEE Conference on Computer Vision and Pattern Recognition (CVPR 2018), Salt Lake City, UT, USA, 2018.

[69] M. A. de Jesus, and V. V. Estrela, Optical flow estimation using total least squares variants, Oriental Journal of Computer Science and Technology, 10, 563-579, 2017. doi: 10.13005/ojcst/10.03.03.

[70] A. M. Coelho, and V. V. Estrela, State-of-the-art motion estimation in the context of 3DTV,

Ed. ;ReubenA. Farrugia,CarlJ. Debono. (Ed.). MultimediaNetworking and Coding. 1ed., IGI Global,Hershey,PA,USA,2013,pp. 148 – 173,ISBN:1466626607.

[71] V. V. Estrela,M. H,da Silva Bassani,and L. A. Rivera,Expectation – Maximization Technique and Spatial – Adaptation Applied to Pel – Recursive Motion Estimation,Proceedings of the WSCG 2004,Pilzen,Czech Republic,2004.

[72] R. Liao,X. Tao,R. Li,Z. Ma,and J. Jia,Video super – resolution via deep draft – ensemble learning,in Proc ICCV,Santiago,Chile,2015.

[73] A. Kappeler,S. Yoo,Q. Dai,and A. K. Katsaggelos,Video superresolution with convolutional neural networks,IEEE Transactions on Computational Imaging,2,2,109 – 122,2016.

[74] A. Kappeler,S. Yoo,Q. Dai,and A. K. Katsaggelos,Super – resolution of compressed videos using convolutional neural networks,in Proc ICIP 2016,Phoenix,Arizona,USA,2016.

[75] V. V. Estrela,H. A. Magalhaes,and O. Saotome,Total variation applications in computer vision. In N. K. Kamila (Ed) Handbook of Research on Emerging Perspectives in Intelligent Pattern Recognition,Analysis,and Image Processing,pp. 41 – 64. IGI Global,Hershey,PA,USA, 2016. DOI:10. 4018/978 – 1 – 4666 – 8654 – 0. ch002.

[76] D. Liu,Z. Wang,Y. Fan,et al.,Robust video super – resolution with learned temporal dynamics,in Proc 2017 IEEE International Conference on Computer Vision (ICCV),Venice, Italy,2017.

[77] D. Liu,Z. Wang,Y. Fan,et al.,Learning temporal dynamics for video superresolution:A deep learning approach,IEEE Transactions on Image Processing,27(7),3432 – 3445,2018.

[78] J. Caballero,C. Ledig,A. P. Aitken,et al.,Real – time video super – resolution with spatiotemporal networks and motion compensation,in Proc 2017 IEEE Conference on Computer Vision and Pattern Recognition (CVPR 2017),Honolulu,Hawaii,USA,2017.

[79] M. Jaderberg,K. Simonyan,A. Zisserman,and K. Kavukcuoglu,Spatial transformer networks,in Proc Twenty – ninth Conference on Neural Information Processing Systems (NIPS 2015), Montreal,Canada,2015.

[80] X. Tao,H. Gao,R. Liao,J. Wang,and J. Jia,Detail – revealing deep video super – resolution,in Proc. 2017 International Conference on Computer Vision (ICCV 2017),Venice,Italy,2017.

[81] Y. Huang,W. Wang,and L. Wang Video super – resolution via bidirectional recurrent convolutional networks,TPAMI,40,4,1015 – 1028,2018.

[82] J. Guo,and H. Chao,Building an end – to – end spatial – temporal convolutional network for video super – resolution,in Proc Thirty – First AAAI Conference on Artificial Intelligence (AAAI – 17),San Francisco,California,USA,2017.

[83] A. Graves,S. Fernandez,and J. Schmidhuber,Bidirectional LSTM networks for improved phoneme classification and recognition,in Proc International Conference on Artificial Neural Networks 2005 (ICANN 2005),Warsaw,Poland,2005.

[84] M. S. Sajjadi,R. Vemulapalli,and M. Brown,Frame – recurrent video super – resolution,in Proc. 2018 IEEE Conference on Computer Vision and Pattern Recognition (CVPR 2018),Salt Lake City,UT,USA,2018.

第9章 深度学习作为无人机系统中超分辨率成像的可替代方案

[85] Z. Zhang, and V. Sze, Fast: a framework to accelerate superresolution pro – cessing on compressed videos, in 2017 IEEE Conference on Computer Vision and Pattern Recognition Workshops (CVPRW 2017), Honolulu, Hawaii, USA, 2017.

[86] Y. Jo, S. W. Oh, J. Kang, and S. J. Kim, Deep video superresolution network using dynamic upsampling filters without explicit motion compensation, in 2018 IEEE Conference on Computer Vision and Pattern Recognition (CVPR 2018), Salt Lake City, UT, USA, 2018.

[87] S Vasu, T. M. Nimisha, and N. A. Rajagopalan, Analyzing perception – distortion tradeoff using enhanced perceptual super – resolution network. Proc, European Conference on Computer Vision (ECCV) workshops, Munich, Germany, 2018, 114 – 131.

[88] Y. Yuan, S. Liu, J. Zhang, Y. Zhang, C. Dong, and L. Lin, Unsupervised image super – resolution using cycle – in – cycle generative adversarial networks, in 2018 IEEE Conference on Computer Vision and Pattern Recognition Workshop (CVPRW 2018), Salt Lake City, UT, USA, 2018.

[89] A. Bulat, J. Yang, and G. Tzimiropoulos, To learn image superresolution, use a gan to learn how to do image degradation first, in Proceedings of the European Conference on Computer Vision (ECCV), Munich, Germany, 2018, pp. 185 – 200.

[90] C. Ledig, L. Theis, F. Huszar, et al., Photo – realistic single image super – resolution using a generative adversarial network. In Proc. 2017 IEEE Conference on Computer Vision and Pattern Recognition (CVPR), Honolulu, Hawaii, USA, 2017, pp. 105 – 114.

[91] Y. Huang, J. Li, X. Gao, L. He, and W. Lu, Single image super – resolution via multiple mixture prior models, IEEE Transactions on Image Processing, 27, 5904 – 5917, 2018.

[92] A. M. Coelho, and V. V. Estrela, Data – driven motion estimation with spatial adaptation. International Journal of Image Processing (IJIP), 6(1), 54, 2012.

[93] V. V. Estrela, H. A. Magalhaes, and O. Saotome, Total variation applications in computer vision, Ed.: Narendra Kumar Kamila, Handbook of Research on Emerging Perspectives in Intelligent Pattern Recognition, Analysis, and Image Processing. IGI Global, Hershey, PA, USA, 2016. doi: 10.4018/978 – 1 – 4666 – 8654 – 0.ch002, http://www.igiglobal.com/chapter/total – variation – appli – cations – in – computer – vision/141626

[94] M. Elad, and A. Feuer, Restoration of a single superresolution image from several blurred, noisy, and undersampled measured images. IEEE Transactions on Image Processing, 6(12), 1646 – 1658, 1997.

[95] D. Rajan, and S. Chaudhuri, An MRF – based approach to generation of super – resolution images from blurred observations, Journal of Mathematical Imaging and Vision, 16, 5 – 15, 2002.

[96] A. Kanemura, S. – i. Maeda, and S. Ishii, Superresolution with compound Markov random fields via the variational EM algorithm. Neural Networks, 22(7), 1025 – 1034, 2009.

[97] K. V. Suresh, G. M. Kumar, and A. Rajagopalan, Superresolution of license plates in real traffic videos, IEEE Transactions on Intelligent Transportation Systems, 8(2), 321 – 331, 2007.

[98] W. Zeng, and X. Lu, A generalized DAMRF image modeling for super – resolution of license plates. IEEE Transactions on Intelligent Transportation Systems, 13(2), 828 – 837, 2012.

[99] S. Mallat, and G. Yu, Super – resolution with sparse mixing estimators. IEEE Transactions on

Image Processing,19(11),2889 - 2900,2010.

[100] W. Dong, D. Zhang, G. Shi, and X. Wu, Image deblurring and super - resolution by adaptive sparse domain selection and adaptive regularization. IEEE Transactions on Image Processing, 20(7),1838 - 1857,2011.

[101] A. Marquina, and S. J. Osher, Image super - resolution by TV - regularization and Bregman iteration, Journal of Scientific Computing,37(3),367 - 382,2008.

[102] M. K. Ng, H. Shen, E. Y. Lam, and L. Zhang, A total variation regularization based super - resolution reconstruction algorithm for digital video, EURASIP Journal on Advances in Signal Processing,2007. doi: 10. 1155/2007/74585.

[103] S. Farsiu, M. D. Robinson, M. Elad, and P. Milanfar, Fast and robust multiframe super resolution. IEEE Transactions on Image Processing,13(10),1327 - 1344,2004.

[104] L. I. Rudin, S. Osher, and E. Fatemi, Nonlinear total variation based noise removal algorithms, Physica D: Nonlinear Phenomena,60,259 - 268,1992.

[105] F. Knoll, K. Bredies, T. Pock, and R. Stollberger, Second order total gen - eralized variation (TGV) for MRI, Magnetic Resonance in Medicine,65(2),480 - 491,2011. 38. P. Getreuer, Total variation inpainting using split Bregman, Image Processing On Line,2,147 - 157,2012.

[106] A. Deshpande, and P. P. Patavardhan, super - resolution of Long Range Captured Iris Image Using Deep Convolutional Network, IOS Series: Advances in Parallel Computing. Volume 31, Published 2017, Netherlands.

[107] V. V. Estrela, O. Saotome, M. A. de Jesus, and D. Stutz, Super - resolution via particle swarm optimization variants, biologically rationalized computing techniques for image processing applications, Eds. : Valentina E. Balas, Jude Hemanth, Berlin: Springer,2017, pp. 317 - 337, (book chapter). doi: 10. 1007/ 978 - 3 - 319 - 61316 - 1_14.

[108] A. Bini, and M. Bhat, A nonlinear level set model for image deblurring and denoising, The Visual Computer,30(3),311 - 325,2014.

[109] D. Cai, K. Chen, Y. Qian, and J. - K. Kamarainen, Convolutional low - resolution fine - grained classification, Pattern Recognition Letters,2017.

[110] J. Li, X. Liang, Y. Wei, T. Xu, J. Feng, and S. Yan, Perceptual generative adversarial networks for small object detection, in In Proc. 2017 IEEE Conference on Computer Vision and Pattern Recognition (CVPR), Honolulu, Hawaii, USA,2017.

[111] W. Tan, B. Yan, and B. Bare, Feature super - resolution: make machine see more clearly, in Proc. 2018 IEEE Conference on Computer Vision and Pattern Recognition (CVPR 2018), Salt Lake City, UT, USA,2018.

[112] N. Huang, Y. Yang, J. Liu, X. Gu, and H. Cai, Single - image super - resolution for remote sensing data using deep residual - learning neural network, in Proc. 24th International Conference on Neural Information Processing (ICONIP 2017), Guangzhou, China,2017.

[113] D. S. Jeon, S. - H. Baek, I. Choi, and M. H. Kim, Enhancing the spatial resolution of stereo images using a parallax prior, in Proc. 2018 IEEE Conference on Computer Vision and Pattern Recognition (CVPR 2018), Salt Lake City, UT, USA,2018.

[114] R. Timofte, E. Agustsson, L. Van Gool, et al., Ntire 2017 challenge on single image super-resolution: methods and results, in Proc. 2017 IEEE Conference on Computer Vision and Pattern Recognition Workshop (CVPRW 2017), Salt Lake City, UT, USA, Honolulu, Hawaii, USA, 2017.

[115] B. J. Maiseli, O. A. Elisha, and H. Gao, A multi-frame super-resolution method based on the variable-exponent nonlinear diffusion regularizer, EURASIP Journal on Image and Video Processing, 2015:22, 2015. doi: 10.1186/s13640-015-0077-2.

[116] S. Durand, J. Fadili, and M. Nikolova, Multiplicative noise removal using L1 fidelity on frame coefficients, Journal of Mathematical Imaging and Vision, 36(3), 201-226, 2010.

[117] S. Yin, X. Li, H. Gao, and O. Kaynak, Data-based techniques focused on modern industry: an overview, IEEE Transactions on Industrial Electronics, 62, 657-667, 2015.

[118] S. Yin, S. X. Ding, X. Xie, and H. Luo, A review on basic data-driven approaches for industrial process monitoring, IEEE Transactions on Industrial Electronics, 61(11), 6418-6428, 2014.

[119] W. Zeng, X. Lu, and S. Fei, Image super-resolution employing a spatial adaptive prior model, Neurocomputing, 162, 218-233, 2015.

[120] A. M. Coelho, and V. V. Estrela, EM-based mixture models applied to video event detection. Ed.: Parinya Sanguansat, In Principal Component Analysis-Engineering Applications, IntechOpen, London, UK, 2012, 101-124.

[121] Z. Ren, C. He, and Q. Zhang, Fractional order total variation regularization for image super-resolution, Signal Processing, 93(9), 2408-2421, 2013.

[122] Q. Yuan, L. Zhang, and H. Shen, Multiframe super-resolution employing a spatially weighted total variation model, IEEE Transactions on Circuits and Systems for Video Technology, 22(3), 379-392, 2012.

[123] A. Shocher, N. Cohen, and M. Irani, Zero-shot super-resolution using deep internal learning, in Proc. 2018 IEEE Conference on Computer Vision and Pattern Recognition (CVPR 2018), Salt Lake City, UT, USA, 2018.

[124] K. Zhang, W. Zuo, and L. Zhang, Learning a single convolutional super-resolution network for multiple degradations, in Proc CVPR, 2018.

[125] T. Jiang, Y. Zhang, X. Wu, Y. Rao, and M. Zhou, Single Image super-resolution via Squeeze and Excitation Network. In Proc. 2018 British Machine Vision Conference (BMVC 2018), Newcastle, UK, 2018.

[126] K. Zeng, J. Yu, R. Wang, C. Li, and D. Tao, Coupled deep autoencoder for single image super-resolution, IEEE Transactions on Cybernetics, 47, 27-37, 2017.

[127] D. Xu, X. Alameda-Pineda, J. Song, E. Ricci, and N. Sebe, Cross-Paced Representation Learning with Partial Curricula for Sketch-based Image Retrieval, arXiv 2018.

[128] G. Riegler, A. O. Ulusoy, and A. Geiger. OctNet: Learning Deep 3D Representations at High Resolutions. In Proc. 2017 IEEE Conference on Computer Vision and Pattern Recognition (CVPR), 2017, Honolulu, Hawaii, USA, pp. 6620-6629.

[129] P. Lindstrom, and J. D. Cohen, On-the-fly decompression and rendering of multiresolution terrain. In Proc. 2010 Symposium on Interactive 3D Graphics (SI3D 2010), Washington, DC, USA, 2010.

[130] D. Ulyanov, A. Vedaldi, and V. Lempitsky, Deep image prior, in in Proc. 2018 IEEE Conference on Computer Vision and Pattern Recognition (CVPR 2018), Salt Lake City, UT, USA, 2018.

[131] W. Dong, Z. Yan, X. Li, and G. Shi, Learning hybrid sparsity prior for image restoration: Where deep learning meets sparse coding, arXiv preprint arXiv:1807.06920, 2018.

[132] Y. Blau, R. Mechrez, R. Timofte, T. Michaeli, and L. Zelnik-Manor, The 2018 pirm challenge on perceptual image super-resolution, in Proc. European Conference on Computer Vision (ECCV 2018), Munich, Germany, 2018.

[133] I. Kassamakov, A. Nolvi, and E. Haggstrom, 3D super-resolution label-free imaging. 2016 Conference on Lasers and Electro-Optics (CLEO), SPIE, San Jose, CA, USA, 2016, pp. 1-2.

[134] K. Jiang, Z. Wang, P. Yi, J. Jiang, J. Xiao, and Y. Yao, Deep distillation recursive network for remote sensing imagery super-resolution. Remote Sensing, 10, 1700, 2018.

[135] K. I. Kim, and Y. Kwon, Single-image super-resolution using sparse regression and natural image prior, TPAMI, 32, 6, 1127-1133, 2010.

[136] B. Lim, S. Son, H. Kim, S. Nah, and K. M. Lee, Enhanced deep residual networks for single image super-resolution, In Proc. 2017 IEEE Conference on Computer Vision and Pattern Recognition Workshop (CVPRW 2017), Honolulu, Hawaii, USA, 2017.

[137] E. Agustsson, and R. Timofte, Ntire 2017 challenge on single image super-resolution: dataset and study, In Proc. 2017 IEEE Conference on Computer Vision and Pattern Recognition Workshop (CVPRW 2017), Honolulu, Hawaii, USA, 2017.

[138] M. Haris, G. Shakhnarovich, and N. Ukita, Deep back-projection networks for super-resolution, in Proc. 2018 IEEE Conference on Computer Vision and Pattern Recognition (CVPR 2018), Salt Lake City, UT, USA, 2018.

[139] C. Ancuti, C. O. Ancuti, R. Timofte, et al., Ntire 2018 challenge on image dehazing: methods and results, in Proc. 2018 IEEE Conference on Computer Vision and Pattern Recognition Workshop (CVPRW 2018), Salt Lake City, UT, USA, 2018.

[140] H. R. Marins, and V. V. Estrela, On the Use of Motion Vectors for 2D and 3D Error Concealment in H.264/AVC Video. Feature Detectors and Motion Detection in Video Processing. IGI Global, Hershey, PA, USA, 2017, pp. 164-186. doi:10.4018/978-1-5225-1025-3.ch008.

[141] E. Hidago-Pena, L. F. Marin-Urias, F. Montes-Gonzalez, A. Marin-Hernandez, and H. V. Rios-Figueroa, Learning from the Web: recognition method based on object appearance from Internet images, in Proc. 2013 ACM/IEEE Int'l Conference on Human-Robot Interaction. IEEE, Tokyo, Japan, 2013, pp. 139-140.

[142] J. Wang, S. Krishnan, M. J. Franklin, K. Goldberg, T. Kraska, and T. Milo, A Sample-and-Clean Framework for Fast and Accurate Query Processing on Dirty Data, in Proc. ACM SIGMOD Int'l Conf Management of Data, Snowbird, UT, USA, 2014.

[143] R. Galland, G. Grenci, A. Aravind, V. Viasnoff, V. Studer, and J. Sibarita, 3D high – and super – resolution imaging using single – objective SPIM, Nature Methods, 12, 641 – 644, 2015.

[144] S. Park, W. Kang, Y. H. Kwon, et al., Superresolution fluorescence microscopy for 3D reconstruction of thick samples, Molecular Brain, 11(1), 17, 2018.

[145] Z. Wang, A. C. Bovik, H. R. Sheikh, and E. P. Simoncelli, Image quality assessment: from error visibility to structural similarity, TIP, 13, 4, 600 – 612, 2004.

[146] Z. Wang, E. P. Simoncelli, and A. C. Bovik, Multiscale structural similarity for image quality assessment, in Proc. The Thirty – Seventh Asilomar Conference on Signals, Systems & Computers 2003 (2003 ASILOMAR), Pacific Grove, CA, USA, 2003.

[147] R. Zhang, P. Isola, A. A. Efros, E. Shechtman, and O. Wang, The unrea – sonable effectiveness of deep features as a perceptual metric, in Proc. 2018 IEEE Conference on Computer Vision and Pattern Recognition (CVPR 2018), Salt Lake City, UT, USA, 2018.

[148] E. Prashnani, H. Cai, Y. Mostofi, and P. Sen, PieAPP: perceptual image – error assessment through pairwise preference, in Proc. 2018 IEEE Conference on Computer Vision and Pattern Recognition (CVPR 2018), Salt Lake City, UT, USA, 2018.

[149] S. Anwar, S. Khan, and N. Barnes, A deep journey into super – resolution: a survey, CoRR, 2019, abs/1904.07523.

[150] E. A. Ogada, Z. Guo, and B. Wu, An alternative variational framework for image denoising, in Abstract and Applied Analysis, Volume 2014 (Hindawi Publishing Corporation 410 Park Avenue 15th Floor, #287 pmb New York, NY 10022 USA, 2014).

[151] P. Perona, and J. Malik, Scale – space and edge detection using anisotropic diffusion, IEEE Transactions on Pattern Analysis and Machine Intelligence, 12(7), 629 – 639, 1990.

[152] Z. Guo, J. Sun, D. Zhang, and B. Wu, Adaptive Perona – Malik model based on the variable exponent for image denoising, IEEE Transactions on Image Processing, 21(3), 958 – 967, 2012.

[153] S. Levine, Y. Chen, and J. Stanich, Image restoration via nonstandard diffusion. Duquesne University, Department of Mathematics and Computer Science Technical Report: 04 – 01, 2004.

[154] J. Weickert, Anisotropic diffusion in image processing, Volume 1. (Teubner Stuttgart, 1998).

[155] R. Courant, K. Friedrichs, and H. Lewy, On the partial difference equations of mathematical physics, IBM Journal of Research and Development, 11(2), 215 – 234, 1967.

[156] D. Giordan, Use of unmanned aerial vehicles in monitoring application and management of natural hazards, Journal Geomatics, Natural Hazards and Risk, 8, 2017.

[157] K. Anderson, and K. J. Gaston, Lightweight unmanned aerial vehicles will revolutionize spatial ecology, Frontiers in Ecology and the Environment, 11(3), 138, 2013.

[158] S. Getzin, R. S. Nuske, and K. Wiegand, Using unmanned aerial vehicles (UAS) to quantify spatial gap patterns in forests, Remote Sensing, 6(8), 6988 – 7004, 2014.

[159] D. Giordan, A. Manconi, F. Remondino, and F. Nex, Use of unmanned aerial vehicles in monitoring application and management of natural hazards, Geomatics, Natural Hazards and Risk, 8, 1 – 4, 2017.

[160] J. Yang, J. Wright, T. S. Huang, and Y. Ma, Image super – resolution via sparse representation, IEEE Transactions on Image Processing, 19(11), 2861 – 2873, 2010.

[161] R. Zeyde, M. Elad, and M. Protter, On single image scale – up using sparse – representations, in *Curves Surfaces*. Berlin, Germany: Springer, 2012, pp. 711 – 719.

[162] M. Haris, K. Sawase, M. R. Widyanto, and H. Nobuhara, An efficient super resolution based on image dimensionality reduction using accumulative intensity gradient, Journal of Advanced Computational Intelligence and Intelligent Informatics, 18(4), 518 – 528, 2014.

[163] S. C. Park, M. KyuPark, and M. G. Kang, super – resolution Image Reconstruction: A Technical Overview, IEEE Signal Processing Magazine, 20, 3, 21 – 6, May 2003.

[164] H. Jia, A new cross diamond search algorithm for block motion estimation, In Proc. IEEE International Conference on Acoustics, Speech and Signal Processing, Montreal, Quebec, Canada, 2004.

[165] Y. Huang, W. Wang, and L. Wang, Bidirectional recurrent convolutional networks for multi – frame super – resolution, in Proc NIPS, Montreal, Quebec, Canada, 2015.

[166] W. – S. Lai, J. – B. Huang, N. Ahuja, and M. – H. Yang, Deep Laplacian pyramid networks for fast and accurate superresolution, in Proc. 2017 IEEE Conference on Computer Vision and Pattern Recognition (CVPR 2018), Honolulu, Hawaii, USA, 2017.

[167] M. Ciocarlie, C. Pantofaru, K. Hsiao, G. Bradski, P. Brook, and E. Dreyfuss, A side of data with my robot, IEEE Robotics and Automation Magazine, 18(2), 44 – 57, 2011.

[168] H. Y. Chow, K. S. Lui, and E. Y. Lam, Efficient on – demand image trans – mission in visual sensor networks, EURASIP Journal on Advances in Signal Processing, 1, 1 – 11, 2007.

[169] B. Liu, Y. Chen, E. Blasch, and K. Pham, A Holistic Cloud – Enabled Robotics System for Real – Time Video Tracking Application, in Future Information Technology, 2014, pp. 455 – 468.

[170] A. A. Proia, D. Simshaw, and K. Hauser, Consumer cloud robotics and the fair information practice principles: Recognizing the challenges and opportunities ahead, Minnesota Journal of Law, Science & Technology, 2014.

[171] T. Winkler, and B. Rinner, Security and privacy protection in visual sensor networks: a survey, ACM Computing Surveys, 47(1):1 – 42, 2014.

[172] K. Ren, C. Wang, and Q. Wang, Security challenges for the public cloud, IEEE Internet Computing, 16(1), 69 – 73, 2012.

[173] W. Beksi, and N. Papanikolopoulos, Point cloud culling for robot vision tasks under communication constraints, in International Conference on Intelligent Robots and Systems (IROS), Chicago, IL, USA, 2014.

[174] K. Lai, and D. Fox, Object recognition in 3D point clouds using web data and domain adaptation, International Journal of Robotics Research (IJRR), 29(8), 1019 – 1037, 2010.

[175] C. Aguero, N. Koenig, I. Chen, and G. A. Pratt, Inside the Virtual Robotics Challenge: Simulating Real – time Robotic Disaster Response, IEEE Transactions on Automation Science and Engineering (T – ASE): SI on Cloud Robotics and Automation, 12(2), 2015, 494 – 506.

[176] K. E. Bekris, R. Shome, A. Krontiris, and A. Dobson, Cloud Automation: Precomputing Roadmaps for Flexible Manipulation, IEEE Robotics & Automation Magazine: SI on Emerging Advances and Applications in Automation, 2015;22:41-50.

[177] L. Wang, M. Liu, and M. Q. -H Meng, Real-time multi-sensor data retrieval for cloud robotic systems, IEEE Transactions on Automation Science and Engineering (T-ASE): Special Issue on Cloud Robotics and Automation, 12(2), 2015, 507-518.

[178] S. Jordán, T. Haidegger, L. Kovács, I. Felde, and I. J. Rudas, The rising pro-spects of cloud robotic applications. 2013 IEEE 9th International Conference on Computational Cybernetics (ICCC), Tihany, Hungary, 2013;327-332.

[179] A. Kasper, Z. Xue, and R. Dillmann, The KIT object models database: an object model database for object recognition, localization and manipulation in service robotics, International Journal of Robotics Research (IJRR), 31(8), 927-934, 2012.

[180] J. Mahler, S. Krishnan, M. Laskey, et al., Learning accurate kinematic control of cable-driven surgical robots using data cleaning and Gaussian process regression, in IEEE Int'l Conference on Automation Science and Engineering (CASE), Taipei, Taiwan, 2014.

[181] G. Mohanarajah, D. Hunziker, M. Waibel, and R. D'Andrea, Rapyuta: a cloud robotics platform, IEEE Transactions on Automation Science and Engineering (T-ASE), 12, 1-13, 2014.

[182] G. Mohanarajah, V. Usenko, M. Singh, M. Waibel, and R. D'Andrea, Cloud-based collaborative 3D mapping in real-time with low-cost robots, IEEE Transactions on Automation Science and Engineering (T-ASE): Special Issue on Cloud Robotics and Automation, 12(2), 2015, 423-431.

[183] L. Riazuelo, J. Civera, and J. Montiel. C2TAM: A Cloud Framework for Cooperative Tracking and Mapping, Robotics and Autonomous Systems, 62(4), 401-413, 2013.

[184] L. Riazuelo, M. Tenorth, D. Marco, et al., RoboEarth semantic mapping: a cloud enabled knowledge-based approach, IEEE Transactions on Automation Science and Engineering (T-ASE): Special Issue on Cloud Robotics and Automation, 12(2), 2015, 432-443.

[185] J. Salmeron-Garcia, F. Diaz-del Rio, P. Inigo-Blasco, and D. Cagigas, A trade-off analysis of a cloud-based robot navigation assistant using stereo image processing, IEEE Trans. Automation Science and Engineering (T-ASE): SI on Cloud Robotics and Automation, 12(2), 444-454, 2015.

[186] B. Kehoe, S. Patil, P. Abbeel, and K. Goldberg, A Survey of Research on Cloud Robotics and Automation, IEEE Transactions on Automation Science and Engineering, 12(2), 398-409, 2015.

[187] V. V. Estrela, A. C. B. Monteiro, R. P. França, Y. Iano, A. Khelassi, and N. Razmjooy, Health 4.0: Applications, Management, Technologies and Review. Medical Technologies Journal, 2019;2(4):262-276, DOI: 10.26415/2572-004X-vol2iss1p262-276. 262.

[188] V. V. Estrela, A. Khelassi, A. C. B. Monteiro, et al., Why software-defined radio (SDR) matters in healthcare?. Medical Technologies Journal, vol. 3, no. 3, Oct. 2019, pp. 421-9, doi:10.26415/2572-004X-vol3iss3p421-429.

[189] J. Hemanth, and V. V. Estrela, Deep learning for image processing appli – cations. Adv. Par. Comp. IOS Press, Netherlands. ISBN978 – 1 – 61499 – 821 – 1（print）978 – 1 – 61499822 – 8（online），2017.

[190] N. Razmjooy, and V. V. Estrela, Applications of image processing and soft computing systems in agriculture（pp. 1 – 337）. Hershey, PA: IGI Global, 2019. doi:10. 4018/978 – 1 – 5225 – 8027 – 0.

[191] N. Razmjooy, V. V. Estrela, and H. J. Loschi, A study on metaheuristic – based neural networks for image segmentation purposes, data science theory, analysis and applications, Taylor and Francis, Abingdon, UK, 2019.

[192] N. Razmjooy, M. Ramezani, and V. V. Estrela, A solution for Dubins path problem with uncertainties using world cup optimization and Chebyshev polynomials. In: Iano Y. , Arthur R. , Saotome O. , Vieira Estrela V. , and Loschi H. （Eds）Proceedings of the 4th Brazilian Technology Symposium（BTSym'18）. BTSym 2018, Campinas, SP, Brazil, 2019. Smart Innovation, Systems and Technologies, vol 140. Springer, Cham, Zurich, Switzerland.

第10章 无人机系统的体验质量和服务质量

本章研究了无人机网络-物理系统(UAV-CPS)的服务质量(QoS)和体验质量(QoE)。这些参数有助于收集有关包含飞行节点(FN)、地面站(GS)和其他关联设备在网络中连接选项的数据。它们还有利于解决与子网络选择数量有关的复杂问题,这些问题使设计、容量、频谱利用率、网络覆盖范围和可靠性以及来自飞行 Ad-hoc 网络(FANET)的其他问题变得复杂,特别是在流式系统的情况下。QoS 和 QoE 允许自动发现最佳网络配置和用户应用程序成本。现有的攻击路线按照功能列出。强调限制和要点,以便给出该领域的初始研究和进一步研究。如果 UAV-CPS 网络的 QoS 较低,则实时数据将不准确,并且会遭受数据丢失而导致的不准确情况。数据信息丢失或延迟(由于数据包的丢失、重新排列和延迟)可能会降低 UAV-CPS 用户(运营商)的满意度。QoS 网络的恶化会影响实时视频监控,并且必须始终将其与板载需求一起考虑。这种影响将会导致数据丢失和图像质量降低,从而降低 QoE。

10.1 引言

由于计算组件的固有分布,它们必须应对传感器输入的不确定性,并且需要产生实时响应。具有所有地面和遥感设备的无人机可以视为 CPS,因为它涉及多个计算和物理元素[1-4],这些元素通常在不同类型的网络上进行交互。在本章中,这种系统称为 UAV-CPS。无人机也可以充当 FANET 中的 FN。

CPS 的技术限制带来了一些缺点。UAV-CPS 方案必须解决诸如智能自动驾驶汽车、环境智能、自组织、工厂控制和修复、自我优化、智能电网等问题。UAV-CPS 要求重新考虑用于分析、要求、验证、确认、标准、安全、仿真程序和认证规定的工具以及其他问题。

必须建立可扩展到 CPS 的新解决方案,以便在物理领域和跨学科接口中实现非必要耦合,从而避免干扰。UAV-CPS 在动态环境中运行,由于异常动作、零星活动、开放性或结构变化而导致不确定性。这些未知的依赖性会打乱独立性假设,并要求进行适应。而且,接口和扩展可能会使得方案更加复杂。因此,UAV-CPS 必须具有容错操作机制来应对来自网络或物理层面的恶意攻击。

除此之外,物理成分从质量方面看不同于面向对象的软件成分。因此,依赖于方法调用和线程的标准抽象是不够的。典型的 UAV-CPS 架构如图 10.1 所示。

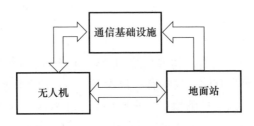

图 10.1　UAV-CPS 架构

QoS 是指 UAV-CPS 在通过网络传输时所具备的最低的丢包率、最大的带宽、最小的延迟、最佳图像或视频处理(如捕获、流传输和记录)[5-6]。QoS 建立在客观参数的基础上,以提高和量化 UAV-CPS 的性能。QoS 是 UAV-CPS 关注的一个重要问题,因为实时监控、捕获和瞄准不同的子系统,涉及高分辨率(HD)和高质量的图像和视频记录以及将这些图像和视频通过网络广播到中央或 GS。

QoE 通过衡量用户的满意度、娱乐性和期望水平来表示最终的用户判断,这是主观指标[7]。它涉及如感知、应用程序的体验或应用程序、信心和网络性能之类的问题。可以从 QoS 当中获得 QoE。

供应商通过 QoE 来调整用户的需求,并通过访谈、基于网络的调查和问卷调查来获取用户有关于产品或服务的主观依据[3,8-13]。但是,UAV-CPS 必须采取整体和连续的 QoE 措施,以使子系统适应必需部分而无需与用户进行过多交互[2]。

除了最终用户对低成本的期望之外,不断增长的网络流量也增加了通过 UAV-CPS 实时监控高质量 QoE 交付对多媒体的重要性。应用层加密的逐步使用、高质量服务的虚拟化网络部署以及用户隐私方面的最新感知调整,促使 QoE 检查解决方案能够适应 UAV-CPS 等快速变化的场景。

10.1.1　从 CPS 角度看机载网络

UAV-CPS 包含一个机载网络(AN),其物理(飞行路径、机动和多模资源,如 GS,地面节点和控制站)与网络(计算、通信和联网)模块之间存在强烈的交互作用(图 10.2)。AN 的最终挑战是通过 FN、网络拓扑结构的重新配置和空中安全多媒体数据共享来促进网络和物理组件之间可信赖的协同交互,从而显著提高无人机姿态感知与 UAV – CPS 的安全性。

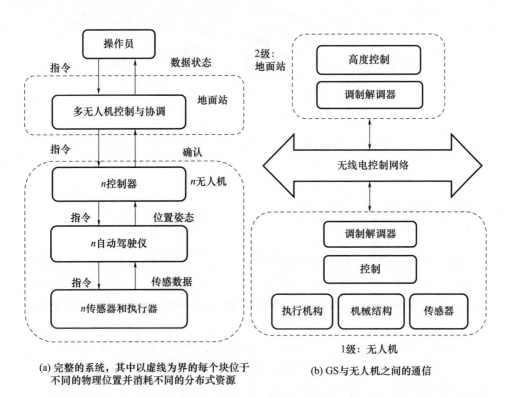

(a) 完整的系统,其中以虚线为界的每个块位于不同的物理位置并消耗不同的分布式资源

(b) GS 与无人机之间的通信

图 10.2　通用 UAV – CPS 网络架构

图 10.3 举例说明了一个由几个空中和地面节点组成的网络。FN 之间的节点高度和距离有很强的可变性。AN 节点可能是静态的也可能是移动的,从而导致 AN 拓扑结构极为动态。高性能 AN 项目必须包含诸如共享信息的完整性和及时性、高带宽提供和可扩展性等因素[14]。

本章的内容如下:10.2 节给出定义;10.3 节给出 UAV – CPS 的应用;10.4 节介绍针对 UAV – CPS 的案例研究;最后在 10.5 节和 10.6 节介绍未来研究趋势、做出讨论和结论。

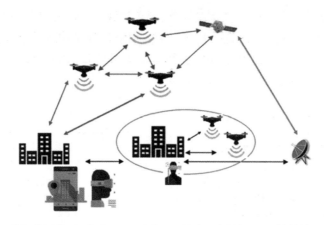

图 10.3 一种机载网络,包含地面、卫星和 RF 链路,可连接至 GS 控制站和其他无人机

10.2 概述

QoS 有助于管理网络基础架构上的数据包丢失、延迟和抖动[5,15]。服务提供商(SP)除了评估和确保质量外,还使用差异化服务和集成服务向用户提供 QoS,但这并不认可 SP 与客户之间的服务级别协议(SLA)[6,16],尽管这两种服务都能非常灵活地访问带宽质量和延迟。

QoS 是 UAV – CPS 与中央数据中心进行通信的一个重要问题,因为它们是一种用于监视远程位置的洪水、冲突、火灾和对象跟踪的实时环境,这些恶劣条件会影响无线接入网(WAN)上的广播。

多媒体流的 QoE 取决于网络损伤和 UAV – CPS 处理视频内容的方式。图 10.4 显示了 WAN 上的端到端多媒体 QoE 原理[17]。多媒体数据在发送方进行编码和打包,而接收方则对数据进行解包和解码。

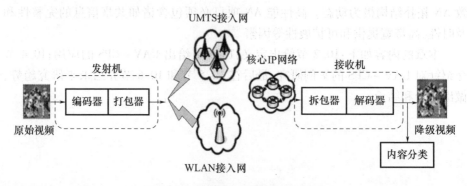

图 10.4 端到端的多媒体 QoE 原理

UAV-CPS 的 QoS 体现了网络的整体性能,包括接入和核心互联网协议(IP)网络。QoE 则依赖于 UAV-CPS 中被测量出的 QoS,如在用户看到的应用场景的实时视频广播中。视频内容的 QoS 是基于视频序列中的不同元素,如视频格式(大小)、实体或信息、帧率和比特率来建立的。

10.2.1 影响 QoS/QoE 的参数

图 10.5 总结了影响 UAV-CPS 中的 QoS 参数。

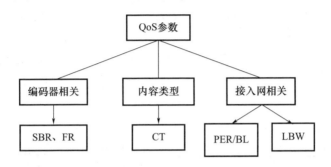

图 10.5　UAV-CPS 中的 QoS 参数

QoS 参数除了内容类型外,还包括应用程序级别和访问网络级别。对网络参数的访问包括数据包丢失、路由器等待和队列延迟、多路径路由数据包重排、链路带宽和平均突发长度等[18-20]。

应用程序级别的参数包括视频解码器、比特率、帧率和传输率,来为用户传递质量信息。此外,UAV-CPS 的视频内容还会影响 QoS,具体取决于:使用的摄像机和用于传输的接入网。大多数的 FN 都具备高质量高清摄像机来拍摄多媒体,高清多媒体传输需要高带宽网络和低流量,但是如果网络资源不可用,则可能发生延迟和多媒体数据丢失(数据包丢失)的情况。

10.2.2 云距离对于 QoS/QoE 的影响

云距离同样也是 UAV-CPS 系统中 QoS/QoE 的首要问题,因为 FN 和云数据中心之间的距离很长,这增加了数据广播中额外的网络延迟[21]。在短距离云和无人飞行器中,通信具有较小的延迟,因为同云和 FN 之间的长距离相比,多个路由器和连接的接口较低,其中不同 SP 的大量路由器和接口会增加额外的延迟[16]。人们青睐无人机高质量的视频和图像,但由于网络带宽窄,云组件和 FN 之间的距离长,他们并没有得到高质量的视频,这就会降低 QoE[22]。像服务器客户端系统一样直接访问 UAV-CPS 是不同的,但是如果 UAV-CPS 是受云控制的,那么影响不同,因为请求是向云管理软件发送的,它将分布在内部机架和集群中,这也增加了数据接收和向 FN 转发命令的内部云延迟。UAV – CPS 和云服务之间的网络距离增加会干扰启动延迟和等待时间,直到服务重新安排以提供所需的 QoE[23-24]。

10.2.3 UAV – CPS 中的 QoS/QoE 监视框架

UAV – CPS 的 QoS/QoE 监控框架基于客观的 QoE,包括:① 从可用 QoS 数据推断 QoE 的客观技术因素;② 与人类生理和认知系统相关的客观人为因素[6]。QoS/QoE 监控框架采用基于技术 QoS 数据所采集的客观 QoE 来评估 QoS 性能,并更好地管理操作。

图 10.6(a)和(b)说明了一种非侵入式评估 UAV – CPS 多媒体的 QoS 和 QoE 的端到端框架。视频被数字化、编码/压缩并发送到接入网和核心 IP 网络。

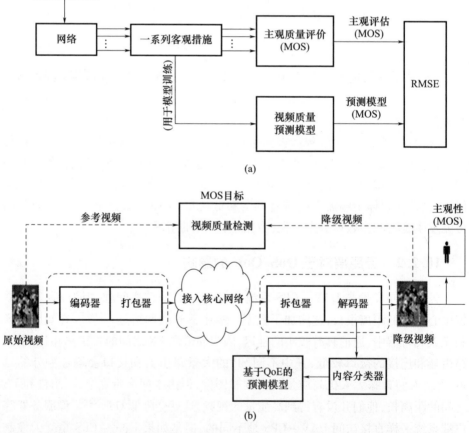

图 10.6　视频质量预测模型框图(a)与端到端的多媒体 QoE/QoS 框架(b)

在接收器处对视频进行解包、解压缩/解码和重构,可以从 VQM、PSNR 等指标中客观地获得 QoS,也可从人的主观意识上通过接收到的视频平均意见分数(MOS)得到 QoS。这些数据有助于开发用于观测 QoS 的非侵入式多媒体模型。MOS 评级见表 10.1。

表 10.1 平均意见得分

MOS	品质	看法
5	极好的	不易察觉的
4	好的	可察觉的
3	公平的	有点烦人的
2	较差的	恼人的
1	差的	非常恼人的

客观 QoE 监控依赖于代理技术。简单网络管理协议(SNMP)[22,25] 提供的应用功能有助于恢复 QoS 数据。该协议有助于从 IP 设备上的服务器、集线器、交换机、打印机和路由器等网络设备收集信息并进行配置。SNMP 使用代理来检索 QoS 网络数据,如从云到 UAV–CPS 的路由知识,除了网络接口数量之外的数据包输入和输出数量。SIGAR[26] 用于低级系统信息,如总的已用内存和实际可用内存,CPU 利用率和具体事项,换句话说就是一个进程所花费的内存和 CPU[27]。图 10.7 给出了一个建议的框架,用于监控应用程序级 QoS(AQoS)和网络级 QoS(NQoS)参数,以从中评估 QoE。对于 UAV–CPS,QoS/QoE 监控软件会监视云环境中是否存在空闲资源,如计算、存储和内部云网络的状态。从云到 UAV–CPS 的 QoS 数据监控包含以下元素。

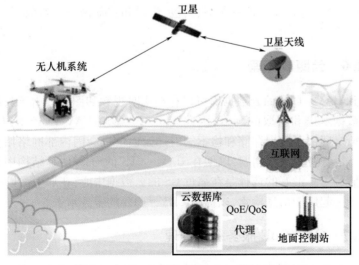

图 10.7 UAV–CPS 中的 QoS/QoE 监视结构

(1)从云到用户的距离；
(2)它们之间的路由器数量；
(3)来自路由器的网络流量和特定延迟；网络带宽；
(4)网络类型；无人机系统能力；
(5)整体系统使用情况；
(6)内存使用情况（CPU 和内存的使用情况对云使用性能具有很大影响）；
(7)路由器队列上的特定延迟；
(8)为管理层了解 UAV-CPS 中 QoS 缺陷所提供的信息。

10.2.4 应用层管理

应用程序级 QoS/QoE 的介入有助于同时正确处理应用程序的同时分析其性能。一种 UAV QoS/QoE 监控结构还包括应用程序级别管理，该管理定期监控基于协议（如应用程序文件传输协议、超文本传输协议、远程桌面协议、实时协议、通用互联网文件系统（CIFS）或 SQL 和交换的应用程序的网络流量。该单元还监控硬件资源利用率以及免费资源和 UAV-CPS 现在执行的各种任务，如图像/视频录制和流媒体。

10.2.5 网络级管理

UAV-CPS 的 QoS/QoE 框架的网络级控制监控 QoS 网络知识，用于在云数据中心和无人机 CPS 部分之间发送和接收数据。代理从控制中心到 UAV-CPS 和云存储，以收集有关 UAV-CPS 和云之间链路容量的数据。代理还测量 UAV-CPS 以及云接口上的网络流量、发送和收集的数据量。这些数据被存储并用于网络分析，查找和分析错误率、丢包、延迟和重新排序等参数，从而有助于正确的网络管理。

10.2.6 云距离管理

连接到 UAV-CPS 的云数据库可以存储实时视频和图像。由于长距离的云层会增加数据发送和获取的延迟，因此云距离会影响 UAV-CPS 的 QoS/QoE。机架和集群之间用于数据存储和检索的内部云通信也会因内部网络堵塞而导致延迟[28-31]，最好避免这些问题，以便在 UAV-CPS 的 QoS/QoE 结构中有一个邻近的云存储位置。最好的选择是在靠近 GS 的云数据中心，在那里可以控制 UAV 或将其与云一起用于信息检索，以提高 UAV-CPS 的服务水平。

10.2.7 QoS/QoE 服务级别管理

QoS/QoE 服务级别管理通过关键性能指标来监视和管理实体的 QoS。传统上,服务级别管理使用传统的监视工具(如 Microsoft SMS),但 UAV – CPS 网络因其组成、动态和灵活以及远程访问服务而受到影响。UAV – CPS 的 QoS/QoE 监控框架使用代理来评估网络带宽、不同时间的峰值利用率和错误率等资源。服务级别管理将实际性能与预定义的期望进行比较,以确定适当的行动并生成有意义的报告[26]。

10.2.8　UAV – CPS 中的 QoS/QoE 指标

已为 UAV – CPS 通信定义了后续的 QoS/QoE 度量标准,以便 QoS/QoE 框架能够由 UAV 操作人员自动测量并报告有关 UAV 部件的激活信息。

(1)吞吐量:对网络中每个流的数据量的广泛度量。它告诉我们在给定的时间内有多少位数据进入网络系统,CPU 如何处理通过内存传输的数据量,操作系统的性能,以及随着网络容量的增长或无线频谱的限制而减少的自组织或移动网络。

(2)网络丢包和延迟:这是在发生数据包丢失时(当数据包被销毁时,它将永远无法恢复),干扰数据从云到 UAV – CPS 的传输的网络流量参数。如果包含输入操作信息的数据包丢失,则操控员将失去对 FN 的控制,从而影响其性能。同样,信息延迟到 UAV – CPS 后到达的数据包将导致发送输入数据包延迟到云端,形成非常规监控。

(3)资源指标:这些指标包含来自 UAV – CPS 硬件资源的日志和报告,如总体系统功能、每个任务的内存使用率和 CPU 利用率、可用资源和资源管理,检查资源将有助于管理和维护 UAV – CPS 的性能。

10.2.9　QoS 到 QoE 的映射

QoS 到 QoE 的映射功能出现在监控系统 AQoS 和 NQoS 上,并使用云距离实现该映射。在 QoS/QoE 无人机监控框架中,QoE 依赖于代理程序捕获的技术 QoS 数据,包括视频、多模态图像、网络和云距离,以及用于可视检索和分析的数据存储知识。如果收集的数据结果存在较长网络延迟或视频流中数据包丢失、机载无人机内存不足或处理能力较低等问题,则需要花费较长时间来处理来自操控员的指令,这会降低 QoS/QoE。这些现象降低了 UAV – CPS 的 QoE,降低了视频/图像记录和传输到控制中心的性能,QoS 参数能够主观或客观地映射到 QoE。

10.2.10　主观与客观测量

主观 QoE 指标利用问卷、访谈、网络调查和投诉箱[32],并参考关于人们所说和经历的指标。此外,与客观 QoE 指标相比,它们成本高、浪费时间且不太准确[33]。有时,受试者是盲目的或无意识的,从而降低了对服务体验的准确评价并提供错误信息。用户从 SLA 中引用的 SP 中获得更多青睐的贪婪性也会引起负面反馈,这也是主观的 QoE 评估警告[34]。

客观的 QoE 指标依赖于技术 QoS 数据和人们对受试者如何完成一项任务的生理测试,而不考虑他们在执行任务时的感受[35]。客观的 QoE 评估比主观评估提供的数据更准确,因为客观的方法是通过基于代理的软件获取数据并记录报告,不需要主体的实时参与即可获得他们的反馈[36]。

对于使用客观或主观指标来决定某件事是否被有意感知,目前尚无共识。主观方法容易出现准则波动。例如,某人可能更倾向于作出反应;他们没有注意到任何取决于人的环境的东西。利用信号检测理论的检测问题和虚警问题,可以部分解决这一问题。然而,多数问题还取决于问题的提法以及如何激发受访者作出反应。是/否类型的问题似乎会自然地融入主观体验,而两种选择的问题则迫使人们选择更具客观性的表现。尽管如此,以更好的方式提出这两类问题可能会影响结果。

10.2.11 衡量 QoS/QoE 的工具

许多众包框架用于收集行业提供的在线 QoS/QoE 工具和研究人员提供的参考模型以评估 QoS/QoE。简单的手工方法用于主观 QoE,如访谈、问卷和投诉箱,以衡量用户的 QoE,并使用 MS Excel 和 Gephi 工具分析收集的数据[37-38]。此外,还开发了先进的自动众包工具,用于在运行环境和数据分析中捕获 QoS/QoE。众包是一种新兴技术,可用于在一个不受控制的环境中衡量最终用户的 QoE。Sajjad 等[39]提供的众包框架可以收集图像和视频的 QoS/QoE,用于衡量最终用户感知的在线视频流的 QoE。该工具还可以实时测量重要的 QoS 网络参数(数据包丢失、延迟、抖动和吞吐量)、检索系统信息(内存、处理能力等)和最终用户系统的其他属性。参考文献[40-44]还提供了用于图像和视频 QoS/QoE 评估的相关众包框架。

供应商提供了客观的 QoE 工具,通过捕获技术 QoS 数据来测量多媒体流的 QoS/QoE[45-46]。Casas 等[47]提供了一种基于机器学习的客观 QoS/QoE 测量模型,该模型能够使用被动式设备内输入作为输入来预测受流行的智能手机应用程序(如 YouTube 和 Facebook)的最终用户体验的 QoE。目标 QoS/QoE 工具也包含在众包框架中,以自动捕获 QoS 技术数据[6,33,41]。

研究人员用参考模型来测量实时环境中的 QoS/QoE,如无参考模型[21,26]、简化参考模型[48]和全参考模型[49]。

除调查人员提供的评估 QoS/QoE 的参考模型外,许多众包工具还用于收集行业设置的在线 QoS/QoE。简单的手工方法测量用户的主观 QoE,获得的数据可以用 MS Excel 和 Gephi 工具进行分析[37,38]。众包是一种在不受控制的情况下推断最终用户在线视频流 QoE 的新兴方法，该工具还可实时评估重要的 QoS 网络参数(如延迟、数据包丢失、抖动和吞吐量)，检索 UAV-CPS 信息(如内存和计算能力)，以及最终用户位置的其他属性[40-44]。先进的自动众包工具还可以在运行环境中以及在执行实时数据分析时捕获 QoS/QoE。

卖方为具有技术 QoS 数据的多媒体流提供了客观的 QoE 工具[45-46]。客观的 QoS/QoE 评估模型在将其作为无源设备内测量的输入时，可能依赖于机器学习来预测对移动应用程序的最终用户(如 YouTube、Twitter 和 Facebook)而言明显的 QoE[47]。众包环境中内置的客观 QoS/QoE 工具用于自动捕获 QoS 技术数据[6,33,41]。

研究人员使用参考模型来测量实时环境中的 QoS/QoE，如使用无参考模型[21,26]、简化参考模型[48]和全参考模型[49]。

(1) 当尝试通过实施检查几个 QoS 参数来预测 QoE 时，无引用模型不知道原始流或源文件。图 10.8 展示了无参考视频质量测量原理。

(2) 简化参考模型利用了关于原始流的有限知识，并试图将其与实时读数相结合以获得 QoE 预测。

(3) 完全参考模型假定可以完全访问参考视频，可能与实时环境测量相结合使用。

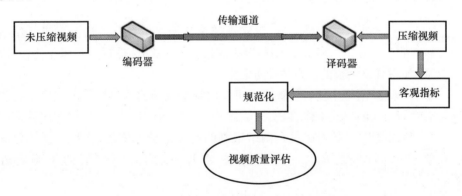

图 10.8　非侵入式(又称无参考)视频质量测量

10.3　应用领域

10.3.1　社交网络、游戏和人机界面

UAV-CPS、社交网络(SN)和娱乐的整合可以产生突破性的影响。例如，

UAV – CPS 视频游戏通过更多来自不同类型传感器的输入增强了对网络世界的视觉感知,从而提高用户对 SN 信息共享的参与度。尽管存在隐私问题,但允许用户调整其隐私设置。参考文献[4,50 – 52]的研究工作讨论了三维远程沉浸式环境中的分布式游戏,该游戏使用了一个概念框架对用户体验的非技术性影响进行建模。社交互动会影响 UAV – CPS 环境中图形界面操控员的 QoE,并可能使用如延迟和视觉质量等的指标,以及诸如年龄等非技术因素。

10.3.2 数据中心

数据中心(DC)可以建模为 CPS。网络部分是负责通信、图像/视频检索和计算的在线应用程序和服务。物理机制允许正确和不间断的操作。网络和物理模块的交互、平衡和时变性使电源管理复杂化。Rao 等[53]提出了利用开/关和动态电压及频率缩放服务器功能来管理数据中心功耗的方法。DC 必须最大限度地利用计算资源来提供服务质量,同时保持计算和冷却的能源成本最小[54 – 55]。DC 模型可以将数据使用与系统架构相关联,以调查网络和物理子系统之间的交互、计算任务的流量和分布,以及物理网络如何处理能源。

10.3.3 电网和能源系统

UAV – CPS 要求新的自由度,以实现遵循智能电网范例的情况下进行供电的可靠节能解决方案。需求子系统必须使用绿色的 WAN 进行多媒体流传输,以应对 WAN 中日益增长的多媒体业务需求。

用于在无线网络中流式传输多个可变比特率视频的功率控制可以集中式和低复杂度分布式的方式工作,以实现最佳调度比特流的广播功率,从而向移动用户和 GS 传送 VBR。必须在复合网络限制的情况下插入模块,并且在保持足够 QoE 要求的同时,避免缓冲区下溢或溢出(收到无线信道的不确定性影响)[56 – 59]。

参考文献[60 – 63]中出现了广域监控系统中的通信路由策略的群集理论模型。对于通信基础设施遭到拒绝服务攻击的情况,模型提供了有效的路由策略来稳定故障电源子系统的瞬变[55,64 – 65]。

10.3.4 网络系统

在许多 UAV – CPS 中,摄像机捕获、发送和接收高分辨率实时静止图像和视频,需要无处不在的宽带网络接入。Xing 等[66 – 68]提出了一种网络架构,用于支持 CPS 中智能手机和 Internet 主机之间的视频通信,这对于需要基于视频的数据采集和共享的多个 UAV – CPS 非常有用。

来自参考文献[69-71]的工作提出了一种技术,该技术可以通过探索节能机会,同时遵守时间和优先级限制,来最大程度帮助减少 UAV-CPS 实时应用的网络能耗。

10.3.5 监视

一个基于互联网连接的 UAV-CPS 警报设计见参考文献[72-74],该警报使用移动通信网络,对 UAV-CPS 恢复力进行定性和定量描述。警报通过感知现实世界的设备将接口终端和操控员连接起来,以将这些数据传输到现实世界的控制和管理端。恢复力表示面向 3S 的设计如下。

（1）稳定性是指 UAV-CPS 即使受到噪声污染或传感器输入的攻击而受到干扰,也能够通过闭环控制,获得稳定的传感驱动性能。

（2）安全性要求系统具有从攻击中恢复/克服攻击的能力。

（3）系统性涉及传感器和执行器的无缝结合。

10.4 实例探究

10.4.1 应用场景 1：UAV-CPS 在交通拥堵管理中的应用

UAV-CPS 必须通过实时交通监控处理交通拥堵监管问题[75-76]。美国联邦航空管理局(FAA)和美国国家航空航天局(NASA)帮助改进无人驾驶飞机系统交通管理(UTM)系统,该系统可控制空中交通拥堵并避免 UAV-CPS 与 FN 发生碰撞。基于云的 UTM 结构将支持操控员在空中飞行时避免所有无人机操作超过低空视线[77],NASA 有许多合作伙伴,他们安排无人机和其他子系统来测试 UTM 技术。同时,NASA 负责适航性、航程和飞行安全,并与其合作伙伴签署了进行 UTM 测试的协议备忘录。美国联邦航空管理局在某些类型的远程空域中测试运行情况[78]。

UTM 将掌握① 有关风和拥堵的信息提供给无人机操控员和数据库,以避免与物体和建筑物发生碰撞；② 有关周围所有物体和建筑物的证据；③ 有关安全飞行的安全地点建议；④ 有关是否飞越机场和交通繁忙路线等限制区域或在其他关键操作期间的紧急情况警报。这些安排将帮助用户选择适当的路径,如图 10.9 所示。

UTM 有针对不同风险情况的四种版本：第一种版本用于在无人区安全飞行和着陆；第二种版本构建出低人口区的飞行控制或安全操作；第三种版本限制了与载人飞机的有限接触；而第四种版本是为了执行城市任务。每一种版本都有

图 10.9　UTM(1英尺=30.48cm)(见彩图)

助于为某些类团体提供具体服务,并支持先前建立的任务或服务。

在多协议环境下,传感器-执行器网络存在许多时间紧迫的问题,如实时监控、实时通信和决策等。多个协议可能有不同的延迟。抖动变化可能会中断实时功能,特别是在通信操作中的事件检测期间。一个需要认真思考的挑战是跨平台的传感器-执行器通信。由于传感器故障对其他传感器和执行器的影响,它对整个 UAV-CPS 的影响很大。它们需要传感器策略来保持冗余性和稳健性。大量采用不同稳健性或安全性协议标准的传感器会大大降低能源效率[3,79]。由于不同的标准具有不同的能源效率,因此多协议环境中的能源效率模型至关重要。另一个独特的挑战是通过可伸缩性来保证能源效率,因为不同的标准具有不同的通信范围和不同数量的智能连接,因此很难找到总体框架的可伸缩性。此外,安全性、信任和信息隐私是新出现的问题,除了协议间通信外,还需要通过不同的协议仔细研究安全性故障。UAV-CPS 的统一框架需要一个具有所有三种组件(即网络、物理和通信组件)的端到端 QoS。

由于网络和物理实体的接口,非正交 QoS 协议有时可能导致病态交互[80]。因此,物理对象和网络对象同样需要新的 QoS 协议。UAV-CPS 的物联网(WoT)结构[1]可以为所有 UAV-CPS、网络、通信和物理元素提供端到端的 QoS,该结构包括以下五层:WoT API、WoT 上下文、WoT 覆盖、WoT 内核和 WoT 设备。这些层有助于构建可轻松连接具有必要 QoS 的网络和物理元素的应用程序[1]。

FN 在 WoT 设备层上互连各种物理设备。一架无人机可处理三个 WoT 层:覆盖层、内核层和设备层。WoT 内核层使用智能调度程序处理 FN 处的 QoS 请求,以通信、事件处理和任务进行优先级排序。

UAV-CPS UCN:①促进 FN 之间的通信;②控制网络组件的 QoS;③从事件流中提取被动上、下文;④使用主动上、下文执行智能控制;⑤与现有决策单元和其他计算功能融合。UAV-CPS UCN 实现了三个 WoT 层:WoT 上下文、API、WoT 覆盖。

10.4.1.1 FN 的 QoS 管理

内核调度程序可以将计算资源分配给具有不同优先级的线程(事件),以满足端到端 QoS 需求。可扩展的抢先调度将通过代码移动策略[8]在计算单元之间分配所有不完整的作业,以动态改变代码片段与其执行位置之间的绑定。WoT 内核还可以检测和分类新连接/断开的物理机制及其相关资源,这对于自配置和即插即用至关重要。WoT 内核将向 WoT 覆盖层和相应的网络对象公开资源规范。

10.4.1.2 UAV – CPS 的 QoS 管理

UAV – CPS UCN 和 WoT 覆盖层提供了应用程序驱动的和网络感知的逻辑抽象,在互联网基础设施上进行,以应对数据、延迟、抖动和带宽等方面的网络不稳定。通过网络资源选择优良和可预测的性能可保证这些功能。UAV – CPS UCN 管理应用程序(网络)和通信级别管理 QoS 需求。由于 UAV – CPS 对每个组件都需要 QoS,因此针对物理对象和网络对象的 QoS 新功能将带大规模替换或调整现有的 TCP/IP,这在当今普及的互联网中是不现实的。因此,在给定 QoS 设置的情况下,WoT 覆盖层在现有 IP 网络的顶部形成了一个应用程序感知的虚拟网络。WoT 覆盖层(包括网络和通信子系统)和 WoT 内核(物理系统)之间的 QoS 处理需要从专用的 QoS 需求到 WoT 覆盖层维护的网络 QoS 参数的接口映射。关于覆盖网络的先前研究以具有数据丢失保证和恢复的更智能的路由[81]为目标[82],WoT 覆盖设置为:① 提供动态模型,将虚拟路径端到端延迟与每个底层 IP 路由器上排队延迟的波动解耦;② 使用网络缓存模型安排多路由、接纳规则。

UAV – CPS UCN 使用以下组件部署 QoS。

(1)覆盖路由器:使用多路径传递来保证时间关键的 RESTful 消息的准时分发。

(2)延迟分析器:采用接纳控制来保证端到端延迟满足 CPS 的 QoS 要求。

(3)网络连接:在主网络中打包/解包 RESTful 消息,同时提供由 QoS 监视器声明的网络特定接口的实现。

(4)QoS 监视器:通过探测获取底层网络的更新条件。

(5)QoS 配置:根据特定于应用程序的需求为 RESTful 消息分配优先级标签。

(6)基于 QoS 的调度器:通过带有优先级标记的优先级队列来规划传入的 RESTful 消息。

10.4.2 应用场景 2:使用智能车辆系统避免拥堵和事故

本节描述了在多无人机环境中处理 QoS 和 QoE 的案例研究。UAV – CPS 包含许多车载无人机节点(IVUAVN)[83]。此外,UAV-CPS 网络(UCN)可以控制 WoT 设备层上的众多传感器和执行器[38,84]。IVUAVN 还可以连接到专用车辆控制器局域网,以便于直接控制 FN 中的各种物理结构,如摄像头、导航和其他关键模块。UCN 在主网络顶部提供 WoT 覆盖,以提供 QoS、背景处理以及运营商、无人机和 UCN 之间的接口。

QoS 要求(包括计算和硬件 QoS)超过了智能 UAV-CPS 的通信要求。如果任何 CPS 组件不能提供 QoS,那么整个系统可能会遇到问题。例如,如果无人机出现意外故障,操控员只有很短的时间来停止它。如果数据传感器快速捕捉到这个事件,但是无人机减慢了一个动作序列,那么可能发生一些错误的操作。

WoT 框架可以通过 UAV-CPS 智能地执行事件通信,同时考虑内部 FN 传感器信息和其他 UAV-CPS 来源的外部数据。例如,如果操控员想要移动无人机,则 UCN 会直接分析来自本地无人机传感器和其他外部节点的读数来做出决策。WAVE 短消息协议有助于将高优先级外部事件快速传输到其他 FN[85]。

UAV-CPS 将通过在 UAV-CPS UCN 上分配一个高优先级来处理 FN 事件消息,然后由计算单元快速决定,因为它具有高优先级事件,如果其优先级比其他任务的优先级更重要,则可能中断其他组件。根据传感器数据,如果结果显示 FN 有危险,则 UAV-CPS 将立即采取行动(如声音警告)。在安全的情况下,操控员可以在没有任何 UAV-CPS 的引导下完成操纵。无人机必须识别所有事件并确定其优先级,以尽量减少对操控员的干扰。事件的优先级排序至关重要,因为它会干扰所有其他 UAV-CPS 模块的 QoS 处理。智能 FN 可减轻 FN 的传感器、执行器和其他机载智能子系统产生的内部中断。例如,机载传感器将不间断地向 UAV-CPS UCN 传输 FN 数据。这些数据在相应的 UAV-CPS 节点中进行优先级排序,具有此优先级的进程将与事件重要性级别一起在 UAV-CPS UCN 排队。如果 FN 操作不稳定且响应时间慢,则 UAV-CPS 将推断操控员无法安全操作无人机。操控员的行为改变将加强事件优先级,从而提高事件消息传递、计算和 UAV-CPS 执行操作的 QoS。在本例中,无人机将向导航系统查询最近的安全位置以停止进程,并建议操控员休息。

无人机内核调度程序处理来自传感器节点的事件信息,以便首先处理关键事件。操控员将看到可能引发事故的重要事件,以便立即采取操作,如有可能,无人机将协助操控员。

由于运行中的无人机会收到许多事件通知,所以向操控员呈现所有数据可能会令人难以处理。因此,操控员将有选择地接收事件消息,以感知可能导致事故的高优先级事件并自动缓解事件。解决方案有两个框架:一个框架带有正常事件(安全情况)的调度器;另一个框架用于可能导致事故的关键事件。来自车内传感器的高优先级事件:首先在 CPS 内核层进行处理,而其他常规事件进程将被搁置；然后这些高优先级事件被发送到 UAV – CPS UCN 进行额外处理。当它从外部源收集高优先级事件时，它的调度程序将中断正常进程。UAV – CPS UCN 还将考虑用户行为，以进一步改进决策。

一旦 UAV – CPS UCN 结束，它将通过命令 FN 通过 WoT 覆盖管理状况，手动（通知操控员立即有危险）、自动（通过使无人机减速）或半自动（UAV – CPS UCN 除了帮助缓解危险外还通知操控员出现危险）。如果无人机无法逃脱碰撞，FN 传感器将检测碰撞点并降低无人机速度。此高优先级事件将由传感器节点处理，停止所有其他进程。UAV – CPS UCN 将根据碰撞严重程度进入高优先级模式，提升 CPS QoS 能力来缓解这种情况，并向其他 FN 发送命令，在请求帮助的同时保护无人机和周围地区的人员。

10.5 未来与挑战

尽管 UAV – CPS 技术不断进步，但在密集大规模应用 UAV – CPS 方面，仍存在以下挑战。

10.5.1 建模和设计

这对 UAV – CPS 来说是一个挑战。硬件描述语言或编程语言不足以对其行为进行建模，UAV – CPS 计算部件的预期性能必须说明其对物理环境的影响。因此，它们需要一种统一的建模方法，具有简单的对接和一致性。此外，真实世界的数据集对于测试和验证新的研究思路至关重要。同时，基于时间和事件的编程语言必须协同工作，以便在不同的时间尺度和空间尺度上有效模拟异步动力学[49]。

在现实世界的 UAV – CPS 中，与传感器和执行器的通信没有完全覆盖当前编程语言[18-19,31]，并发是很自然的，需要适当的抽象来进行直观现实世界的建模。

功能实现了新模块与新设计方法/特性的无缝集成，以最大限度地干扰现有模块，同时扩展 UAV – CPS 的功能。这将在供应商和消费者之间创造新的互动。可扩展的基础设施将支持各种传感器和执行器类型，方便访问大量潜在用户。

验证和确认需要新的方法来验证 UAV – CPS 硬件和软件模块的高度可靠性和可重构性[18,19,31,86]。

性能和时间敏感性意味着正确的短的执行时间会带来高性能。要实现对时间敏感的 UAV – CPS 的要求，需要为分析和综合任务提供高效的抽象和工具。

网络和物理模块相互协调和相互依赖的安全性增加了对故障和攻击的敏感

性。医疗系统和灾难补救等关键任务 UAV – CPS 需要实时、高度安全、可靠操作、自我修复框架,以便使用适当的安全性能指标来判断 UAV – CPS 的安全性,从而实现安全状态检查。

移动性引入了 UAV – CPS 中的速度和同步等进一步的实现问题。UAV – CPS 中的分布式计算和网络控制必须是自适应的,对故障具有稳健性,并在分布信息和本地操作不完整时保持整体态势感知。可扩展性有助于 UAV – CPS 在各种情况下的部署。

功率优化影响 UAV – CPS 的成本效益,是未来机载任务设计中的关键约束,因为计算量的增加会影响能耗。

10.5.2 协作服务

UAV – CPS 服务的两种属性是功能性和非功能性[48]。随着 UAV – CPS 业务的发展,越来越多具有可比功能、竞争力的业务出现,非功能性特征成为服务选择的关键因素,因为它们可以区分具有相同或可比功能的对手 UAV – CPS 业务性能。QoS 已成为当前服务和分布式计算研究的热点领域[10]。在现实世界中,一些依赖于用户的 QoS/QoE 属性(例如吞吐量和响应时间等)的值可能因物理位置、用户、网络条件(如 5G 和 Wi – Fi)和其他客观问题的不同而不同[11]。单个运营商尝试所有候选 UAV – CPS 服务以获得足够的 QoS 数据进行评估是不现实的。因此,预测用户相关的 QoS 值对于运营 UAV – CPS 服务推荐至关重要。

协同过滤(CF)QoS 预测被广泛应用,尤其是邻域 CF[7,12 – 14,28 – 29,80 – 82,85,87 – 91]。基于邻域 CF 的方法主要包括两个阶段:① 发现相似的运营商(或服务)并分析其相似性;② 从相似用户(或服务)产生的历史数据中计算未知的 QoS 值。不过,目前的策略有以下两个主要的复杂因素。

(1)在计算用户(或服务)的相似性时,数据通常非常稀疏。当 QoS 矩阵太稀疏时,来自用户的 QoS 值可能非常稀缺甚至为空。在这种情况下,精确估计用户(或服务)相似度是一个挑战,数据稀疏性会降低预测 QoS 值的准确性。

(2)并不是所有的用户都诚实,他们会提供一些不可靠的 QoS 数据。一些人可能会给出随机或固定的答案,而另一些人(如 SP)可能会声称他们的服务有不错的 QoS 数据,而向他们的竞争对手提供较差的数据[7,82]。

大多数方法在执行 QoS 预测之前不检测和预处理不可靠的 QoS 数据。具有稳健性的 UAV – CPS 服务推荐程序需要处理来自不信任用户的数据。因此,他们要求减轻数据稀疏性的影响,消除冷启动问题[15]和高估计精度。

10.5.3 流

无人机越来越受欢迎,这意味着需要解决与 FN 运动、有限资源以及较高误差率相关的挑战,这证明需要一种自适应工具来增强视频广播的可靠性[92]。自适应前向纠错(FEC)方法可以提高网络传输的视频质量,具有较高的移动自由度。一种自适应视频感知 FEC 机制,使用运动向量[87-88,90-91]细节来更好地实时无人机视频广播,提供更出色的用户体验和更丰富的资源利用。这种机制的回报和缺点可以通过模拟和量化 QoE 指标进行分析和测试。

FANET 的实时视频流可以使用协作通信方式进行实时流式传输,其中 FN 聚集到协作集群中。LTE 系统通过链路将视频数据传输到指定的集群头,采用 IEEE802.11p 在集群内多重传输接收到的视频。与非合作 FN 网络相比,错误隐藏方法和有效的资源分配策略显著改善了接收到的视频质量,提高了 QoE 和 QoS。

10.5.4 安全

UAV-CPS 网络面临多种威胁,可以从包含智能和互联 FN 的无线系统解决方案中受益[93-94]。当协作节点中的恶意行为时,FN 之间的协作可能变得更加复杂。FANET 中安全性和 QoS 之间的调整需要研究以优化协作通信。

同样,基于预防的安全结构可以为逐跳和端到端提供完整性保护和身份验证。另一种选择是采用与 VANET 输出容量、误码率和有效吞吐量类似的安全性和 QoS 设置。

可靠的协作下载和协作转发的安全激励系统可以激励无人机操控员安全地下载和转发数据包[15]。协作下载可以使用与指定验证人签名相连的虚拟检查,以确保安全性和公平性。此外,信誉系统可以激励合作,惩罚恶意节点,鼓励数据包转发并实现可靠性。在整个协作转发阶段,使用汇总的 Camenisch-Lysyanskaya 签名来确保激励机制的安全性。

无人机和以基础设施为中心的指标可以通过对关键邻域(可能是事故威胁)的意识计算来衡量 FN 的安全意识[27]。在计算感知度时,基础设施节点还可以包括每个相邻 FN 的位置误差。这些指标依赖于接收到的多个协作感知消息(CAM)及其安全性、重要性、准确性和无人机方向。在每次传输之前,可以在安全程序中加入椭圆曲线数字签名算法。CAM 位于安全 FIFO 接收器上,以便在轮到它时进行验证。

10.5.5 飞行特设网络

FANET 解决了在没有适当基础设施的情况下需要快速部署网络的情况。FN 可以通过不可预知的网络拓扑修改任意飞行。这种行为要求不断地进行路线调整和重新配置,以保证节点间通信、更广泛的覆盖区域以及 FANET 的自配置和自组织[95-99]。

FN 的移动性和空间布局在路由中也很重要,因此这些路线是习惯性地重新规划以便允许 FN 持续互连。因此,路由必须是动态的,使用有效且直接的协议来增强无人机的自主性,并减少发送方 FN 和目的地 FN 之间的数据传输延迟。

FN 的地理位置在很大程度上影响着它们之间的交流。无人机中继节点将所有采集到的信息统一起来,并将数据中继到控制中心。因此,它的地位是战略性的,以保证良好的网络性能。

在无人机的飞行过程中,传感器比传统的传感器网络更密集地收集信息。因此,他们必须更有效地进行交流,减少它们视线中的障碍。减少覆盖特定区域的 FN 的数量是满足这一要求的可靠方法,尽管天气条件可能会恶化通信(如风和雨)。在传输实时视频数据时,FANET 的负载对信息传输能力提出了更高的要求。到目前为止,还没有专门针对 FANET 的路由协议。

由于 FN 具有较高的移动性,因此更新所有 FN 的位置是一个关键特性,FANET 元素可以采用元启发式模型来获得更好的结果和更高的性能。通过优化通信结构,FANET 场景的未来需求可能会实现,并获得可接受和更精确的结果。其他元启发式优化算法需要测试来验证实践,并批准最佳执行时间算法来实现强制性能,如遗传算法[100-102]、粒子群算法[100-102]、布谷鸟搜索[103]和世界杯[104]等。

FANET 的主要挑战是与覆盖一个区域的多个无人机执行协同探测[96-98]。这一特性导致了一个空中网络,其 FN 在该网络中交换数据。因此,可靠和稳定的通信有助于保持适当的 QoS 和 QoE 水平。大多数 FANET 将来自周围环境的数据汇集在一起,并将它们传输到一个 GS。FANET 拓扑结构有几个 FN 和一个无人机中继来接收来自其他 FN 的原数据并将其传输到 GS。UAV 中继相对于其他 FN 的位置对于保证直接或间接(通过中间设备通信)访问所有其他网络元件的适当网络性能至关重要,这需要更多的研究。

精确的分析模型可减少 FN 和基站之间的信号干扰。这些算法在信道带宽内进行均匀跳频,并通过一些计算智能[89-99]技术在过载或中断单元中获得 QoS 增益,以找到优化的新无人机中继位置[98]。多个主干网的使用扩大了被监测区域的覆盖范围,但需要更多地研究单元的适当位置,以确保网络具有最佳性能。

将移动基站(MBS)数量降至最低,以便通过一组分布式地面终端(GT)实现适当的无线覆盖,这就保证单个 GT 处于至少一个 MBS 的通信范围内。

高效的算法有助于以复杂的、适应性强、灵活的方式操纵无人机。具有连接约束的 FANET 任务规划[105]包括许多参数以及动态变量的交互作用。分布式智能代理框架也有一些算法,在这些算法中,代理自主地组织、协作、协商、决定和触发动作来执行特定的分配。连接问题是很难解决的。多项式时间启发式算法可以帮助解决这个问题,同时通过充分定位无人机来保持高性能的数据传输。FN 必须作为路由器相互通信。

FANET 的一个核心问题是将一个无人机中继放置在其他中继之间以扩展通信。因此,电信和计算智能可以帮助利用 FN – GPS(平均每秒钟发送一次定位数据)和 FN – IMU(以比 GPS 更短的间隔发送定位数据)实时找到更好的位置。

因此,通过覆盖区域布置设备的方式对这种网络构成了挑战,无论是改善还是恶化,它直接影响到网络的性能和移动性。高采样率可以避免这种折中,因为尽管有一些注意事项,但离散控制器的性能还是比连续控制器更好。QoS 探索研究网络资源分配和 UAV – CPS 性能(包括控制器)之间的折中,在 CPS 中的不同离散服务间隔。一般来说,分配更多的网络资源可以提高性能。传统上,网络控制系统试图识别导致网络稳定和高性能的条件,包括传感、控制和驱动。优化策略可以在处理数据包丢失时同时识别通信和控制输入。在保持控制性能的同时,通过事件触发控制可以最大限度地分配网络资源。利用最优控制的一些机制可以创建路径和采样时刻,采用时变采样率控制器,并在增加计算复杂度的情况下获得良好的结果。

10.5.6 用户情绪

随着访问在线视频服务的用户数量增长,流媒体服务往往主导着各种移动流量。由于运营商的观看体验哪怕是很小的提升也会大幅提高盈利能力,因此内容提供商和分销商、网络人员和服务供应商将不得不调整通过 FANET 获得的视频。尽管最近人们正在努力通过更好地利用大数据分析用户的观看行为以改善用户的视频 QoE,这些行为依赖于大数据。但是,在用户观看在线视频时,不可能准确地审查用户的秘密意图和情感状态[106]。因此,在未来需要考虑用户的情绪反应,以获得更好的 QoE 评估。在这种框架下:用户的情绪首先将通过情绪检测环境实时检测[107];然后将执行匹配程序,从情感设计的角度来评估操控员意图和多媒体内容的相似性,以表征 QoE 与几个方面的关系,包括平均比特率、缓冲比和操控员的情感。

10.6 小结

本章研究 UAV – CPS 应用的 QoS 和 QoE。利用涉及 FN、GS 和其他设备的网络中累积的各种连接选项,有助于解决如容量、频谱效率、网络覆盖率和可靠性等复杂问题。任何时候,如选择、调度、切换或路由等问题都需要性能指标、能效和低成本接入。越来越多的网络选择使设计变得复杂。

QoS 和 QoE 允许自动为用户应用程序找到最佳的网络配置和价格。现有攻击路线按功能列出。本章强调了限制和优势,为该领域的进一步研究提供了起点。由于 UAV – CPS 网络设计、FN 路由和 FN 分配问题会影响端到端用户的 QoS/QoE，因此必须不断计算系统输出指标，并重新评估优化目标和任务。此外，由于这些挑战的特点，重要的是要考虑如何使用元启发式方法及其混合方法来延长元启发式的能力，同时解决随机和动态优化问题。

如果 UAV – CPS 网络的 QoS 较低，则实时数据将不准确，并且受信息丢失的影响，从而导致系统不准确。信息丢失或延迟（由于数据包丢失、重新排列和延迟）可能会降低 UAV – CPS 用户（操控员）的满意度。网络 QoS 的降低影响了视频的实时监控。这种影响会导致数据丢失和图像质量降低，从而降低 QoE。

参考文献

［1］ Estrela V. V. , Saotome O. , Loschi H. J. , et al. Emergency Response Cyber – Physical Framework for Landslide Avoidance with Sustainable Electronics, Technologies, 6, 42, 2018. doi: 10. 3390/technologies6020042.

［2］ Kuang L. , Yu L. , Huang L. , et al. A Personalized QoS Prediction Approach for CPS Service Recommendation Based on Reputation and Location – Aware Collaborative Filtering, Sensors, 18, 1556, 2018. doi: 10. 3390/ s18051556.

［3］ Benmir A. , Korichi A. , and Bourouis Alreshoodi M. Survey on QoE/QoS Correlation Models for Video Streaming over Vehicular Ad – hoc Networks, 26（4）, 267 – 287, 2018. doi: 10. 20532/cit. 2018. 1004278.

［4］ Khaitan S. K. , and McCalley J. D. Design Techniques and Applications of Cyber Physical Systems: A Survey, IEEE Systems Journal, 9, 2015, 350 – 365.

［5］ http://docwiki. cisco. com/wiki/Quality_of_Service_Networking.

［6］ Recommendation, I. T. U. T. E. 800: Terms and definitions related to quality of service and network performance including dependability. ITU – T August 1994, 1994.

［7］ ITU – T Report 2007. Definition of Quality of Experience（QoE）, International Telecommunication Union, Liaison Statement, Ref. : TD 109rev2（PLEN/12）, Jan. 2007.

［8］ Fuggetta A. , and Picco G. P. Understanding code mobility, IEEE Transactions on Software Engineering, 24（5）, 342 – 361, 1998.

［9］ Andersen D. , Balakrishnan H. , Kaashoek F. , and Morris R. Resilient Overlay Networks, ACM SIGOPS Operating Systems Review, 35（5）, 145, 2001.

［10］ Subramanian L. , Stoica I. , Balakrishnan H. , and Katz R. OverQoS: an overlay based architecture for enhancing internet QoS, in Proc. 1st Symp. Networked Syst Des and Implem, USA, 2004.

［11］ David G. Big goals get big results, in Proc. 2006 World Business Council for Sustainable Development（WBCSD）Liaison Delegates Meeting, China, 2006. 238 Imaging and sensing for unmanned aircraft systems, volume 2

[12] Intelligent Transportation Systems Committee, IEEE trial – use standard for wireless access in vehicular environments (wave) – networking services, IEEE Std 1609. 3 – 2007, vol. 2007, pp. c1 – 87.

[13] Uzcátegui R. A., and Acosta – Marum G. WAVE: A Tutorial, IEEE Communications Magazine, 47(5), 126 – 133, 2009.

[14] Namuduri K., Wan Y., Gomathisankaran M., and Pendse R. Airborne net – work: a cyber – physical system perspective, in Proc. 1st ACM MobiHoc Workshop on Airborne Networks and Communications (Airborne'12), Hilton Head South Carolina USA, 2012, pp. 55 – 60.

[15] Laghari A. A., and Rehman K. U. On quality of experience (QoE) for multi – media services in communication ecosystem. PhD diss., Institut National des Te′le′ communications, Paris, France, 2012.

[16] Hoang D. D., Paik H. – Y., and Kim C. – K. Service – oriented middleware architectures for cyber – physical systems, International Journal of Computer Science and Network Security, 12 (1), 79 – 87, 2012.

[17] http://insideunmannedsystems.com/nasa – working – to – develop – unmanned – aerial – system – traffic – management – or – utm/.

[18] Johnson T., and Mitra S. Parametrized verification of distributed cyber – physical systems: an aircraft landing protocol case study, in Proc. 2012 IEEE/ACM Third International Conference on Cyber – Physical Systems (2012 ICCPS), Beijing, China, 2012, pp. 161 – 170.

[19] Lee E. Cyber physical systems: design challenges, in Proc. 2008 11th IEEE International Symposium on Object and Component – Oriented Real – Time Distributed Computing (2008 ISORC), Orlando, FL, USA, 2008, pp. 363 – 369.

[20] Baheti R., and Gill H. Cyber – physical systems, The Impact of Control Technology, 2011, pp. 161 – 166.

[21] Laghari A. A., He H., Shafiq M., and Khan A. Impact of storage of mobile on quality of experience (QoE) at user level accessing cloud, in Proc. 2017 IEEE 9th Int'l Conf. on Comm. Soft. and Networks (ICCSN), GuangZhou, China, 2017.

[22] Laghari, K. U. R., and Connelly, K. Toward total quality of experience: A QoE modelling in a communication ecosystem. IEEE Commun. Mag. 50(4), 58 – 65, 2012.

[23] Ferrer A. J., Marques J. M., and Jorba J. Towards the Decentralised Cloud: Survey on Approaches and Challenges for Mobile, Ad hoc, and Edge Computing, ACM Computing Surveys, 51, 111:1 – 111:36, 2019.

[24] Ratasich D., Khalid F., Geissler F., Grosu R., Shafique M., and Bartocci E. A Roadmap Toward the Resilient Internet of Things for Cyber – Physical Systems. IEEE Access, 7, 13260 – 13283, 2019.

[25] Laghari A. A., He H., Shafiq M., and Khan A. Assessing effect of cloud dis – tance on end user's quality of experience (QoE), In Proc. 2016 2nd IEEE Int'l Conf. Computer and Comm. (ICCC), Chengdu, China, 2016, pp. 500 – 505.

[26] https://support.hyperic.com/display/SIGAR/Home

[27] Alhamazani K., Ranjan R., Jayaraman P. P., Mitra K., Wang M., and Huang Z. G. Real-time QoS monitoring for cloud-based big data analytics applications in mobile environments, in Proc. 2014 IEEE 15th Int'l Conf' Mobile Data Management, Brisbane, QLD, Sydney, NSW, Australia, 2014, pp. 337-340.

[28] Estrela V. V., Khelassi, A., Monteiro A. C. B., et al. Why software-defined radio (SDR) matters in healthcare? . Medical Technologies Journal, vol. 3, no. 3, Oct. 2019, pp. 421-9, doi:10.26415/2572-004X-vol3iss3 p421-429

[29] Motlagh N. H., Taleb T., and Osama O., Low-altitude unmanned aerial vehicles-based internet of things services: Comprehensive survey and future perspectives. IEEE Int Things J, 3, p. 899-922, 2016.

[30] Estrela V. V., Monteiro A. C. B., Franc. a R. P., Iano Y., Khelassi A., and Razmjooy N. Health 4.0: Applications, Management, Technologies and Review, Medical Technologies Journal, 2(4), 262-276, 2019. doi: 10.26415/ 2572-004X-vol2iss4p262-276.

[31] Cruz B. F., de Assis J. T., Estrela V. V., and Khelassi A. A Compact SIFT-Based Strategy for Visual Information Retrieval in Large Image Databases, Medical Technologies Journal, 3(2), 402-412, 2019. doi: 10.26415/2572-004X-vol3iss2p402-412.

[32] Wang G., Wang C., Chen A., et al. Service level management using QoS monitoring, diagnostics, and adaptation for networked enterprise systems, In Proc. 2005 9th IEEE Int'l EDOC Enterprise Computing Conference, Enschede, Netherlands, 2005, pp. 239-248.

[33] Laghari A. A., He H., Zardari S., and Shafiq M. Systematic Analysis of Quality of Experience (QoE) Frameworks for Multimedia Services, International Journal of Computer Science and Network Security, 17(5), 121, 2017.

[34] Laghari A. A., Channa M. I., Laghari K. R., Aman M., and Memon M. EQOM: Enhanced Quality of Experience (QoE) Framework for Multimedia Services, UACEE International Journal of Computer Science and its Applications, 3(1), 85-89, 2013.

[35] Laghari A. A., He H., Ibrahim M., and Shaikh S. Automatic Network Policy Change on the Basis of Quality of Experience (QoE), Procedia Computer Science, 107, 657-659, 2017.

[36] Laghari A. A., Rehman K. U., and Connelly K. Toward Total Quality of Experience: A QoE Model in a Communication Ecosystem, IEEE Communications Magazine, 50(4), 2012.

[37] Karim S., Zhang Y., Laghari A. A., and Asif M. R. Real time image proces-sing based drone for detecting and controlling street crimes, in Proc. 17th IEEE Int'l Conf. Comm. Technology, China, 2017.

[38] Khan A., Sun L., and Ifeachor E. QoE Prediction Model and its Application in Video Quality Adaptation over UMTS networks, IEEE Transactions on Multimedia, 14(2), 431-442, 2012.

[39] Khan A., Sun L., and Ifeachor E. C. Content classification based on objective video quality evaluation for MPEG4 video streaming over wireless net-works, Journal of Multimedia, vol. 4, 2009, pp. 228-239.

[40] Mushtaq M. S., Augustin B., and Mellouk A. Crowd-sourcing framework to assess QoE, In Proc. 2014 IEEE Int'l Conf. on Communications (ICC), Sydney, Australia, 2014, pp. 1705-1710.

[41] Ribeiro F. ,Florencio D. ,Zhang C. ,and Seltzer M. Crowdmos: an approach for crowdsourcing mean opinion score studies,In Proc. 2011 IEEE Int'l Conf. Acoustics,Speech and Signal Proc. (ICASSP),Prague,Czech Republic,2011,pp. 2416 - 2419.

[42] Chen K. - T. ,Wu C. - C. ,Chang Y. - C. ,and Lei C - L. A crowdsourceable QoE evaluation framework for multimedia content,In Proc. 17th ACM Int'l Conf. on Multimedia,ACM,New York,NY,USA,2009,pp. 491 - 500.

[43] Keimel C. ,Habigt J. ,Horch C. ,and Diepold K. Qualitycrowd—a framework for crowd - based quality evaluation,In Proc. Picture Coding Symposium (PCS),2012,IEEE,Nagoya,Japan, 2012,2012,pp. 245 - 248.

[44] Rainer B. ,Waltl M. ,and Timmerer C. A web based subjective evaluation platform,In Proc. 2013 Fifth International Workshop on Quality of Multimedia Experience (QoMEX),IEEE,Klagenfurt am Worthersee,Germany,2013,pp. 24 - 25.

[45] Gardlo B. ,Egger S. ,Michael Seufert M. ,and Schatz R. Crowdsourcing 2. 0: enhancing execution speed and reliability of web - based QoE testing,In Proc. 2014 IEEE Int'l Conf. Comm. (ICC),Sydney,Australia,2014,pp. 1070 - 1075.

[46] www. witbe. net

[47] www. qoesystems. com

[48] Han J. Cyber - Physical Systems with Multi - Unmanned Aerial Vehicle - Based Cooperative Source Seeking and Contour Mapping. Ph. D. thesis,Utah State University,2014.

[49] Horvath I. ,and Gerritsen B. Cyber - physical systems: Concepts,technologies and implementation principles,in Tools and Methods of Competitive Engineering Symposium (TMCE),2012, pp. 19 - 36.

[50] Desai K. ,Raghuraman S. ,Jin R. ,and Prabhakaran B. QoE Studies on Interactive 3D Tele - Immersion,2017 IEEE International Symposium on Multimedia (ISM),Taichung,Taiwan, 2017,pp. 130 - 137.

[51] Ning Z. ,Hu X. ,Chen Z. ,et al. A Cooperative Quality - Aware Service Access System for Social Internet of Vehicles,IEEE Internet of Things Journal,5,2506 - 2517,2018.

[52] Chen H. ,Xiong P. ,Gavrilovska A. ,Schwan K. ,and Xu C. A cyber physical integrated system for application performance and energy management in data centers,IGCC,2012,pp. 1 - 10.

[53] Rao L. ,Liu X. ,Xie L. ,and Liu W. Minimizing electricity cost: optimization of distributed internet data centers in a multi - electricity - market environment,Proc. 2010 IEEE INFOCOM, San Diego,CA,USA,2010,pp. 1 - 9.

[54] Deylamsalehi A. ,Cui Y. H. ,Afsharlar P. ,and Vokkarane V. Minimizing Electricity Cost and Emissions in Optical Data Center Networks,IEEE/OSA Journal of Optical Communications and Networking,9,257 - 274,2017.

[55] Parolini L. ,Tolia N. H. ,Sinopoli B. ,and Krogh B. H. A cyber - physical sys - tems approach to energy management in data centers,in Proc. 2010 ICCPS,Stockholm,Sweden,2010,pp. 168 - 177.

[56] Akkaya I. , Liu Y. , and Gorton I. Modeling and analysis of middleware design for streaming power grid applications,in Proc. Industrial Track of the 13th ACM/IFIP/USENIX Int'l Mid-

dleware Conf. , Montreal, Quebec, Canada, 2012.

[57] Cao J. and Li H. Energy – efficient structuralized clustering for sensor based cyber physical systems, in Proc. UIC – ATC. IEEE, Brisbane, QLD, Australia, 2009, pp. 234 – 239.

[58] Li H. , Lai L. , and Poor H. V. Multicast Routing for Decentralized Control of Cyber Physical Systems With An Application in Smart Grid, IEEE Journal on Selected Areas in Communications, 30, 1097 – 1107, 2012.

[59] Lu A. , and Yang G. Input – to – State Stabilizing Control for Cyber – Physical Systems With Multiple Transmission Channels Under Denial of Service. IEEE Transactions on Automatic Control, 63, 1813 – 1820, 2018.

[60] Wei J. , and Kundur D. GOAliE: Goal – Seeking Obstacle and Collision Evasion for Resilient Multicast Routing in Smart Grid, IEEE Transactions on Smart Grid, 7, 567 – 579, 2016.

[61] Kateb R. , Akaber P. , Tushar M. H. K. , Albarakati A. , Debbabi M. , and Assi C. M. Enhancing WAMS Communication Network Against Delay Attacks, IEEE Transactions on Smart Grid, 10, 2738 – 2751, 2019.

[62] Wei J. , and Kundur D. A flocking – based model for DoS – resilient commu – nication routing in smart grid, Proc. 2012 IEEE Global Communications Conference (GLOBECOM), Anaheim, CA, USA, 2012, pp. 3519 – 3524.

[63] Huseinovic A. , Mrdovic S. , Bicakci K. , and Uludag S. A taxonomy of the emerging denial – of – service attacks in the smart grid and countermeasures, In Proc. 2018 26th Telecommunications Forum (TELFOR), Belgrade, Serbia, 2018, pp. 1 – 4.

[64] Sridhar S. , Hahn A. , and Manimaran G. Cyber – Physical System Security for the Electric Power Grid, Proceedings of the IEEE, 100(1), 210 – 224, 2012.

[65] Hong J. , Liu C. , and Govindarasu M. Integrated Anomaly Detection for Cyber Security of the Substations, IEEE Transactions on Smart Grid, 2014.

[66] Guoliang X. , Jia W. , Du Y. , Tso P. , Sha M. , and Liu X. Toward ubiquitous video – based cyber – physical systems, in Proc. IEEE Intl. Conf. on Systems, Man and Cybernetics, Suntec, Singapore, 2008, pp. 48 – 53.

[67] Leitao P. , Colombo A. W. , and Karnouskos S. Industrial Automation Based on Cyber – Physical Systems Technologies: Prototype Implementations and Challenges, Computers in Industry, 81, 11 – 25, 2016. ISSN 0166 – 3615.

[68] Zhang Y. , Qiu M. , Tsai C. , Hassan M. M. , and Alamri A. Health – CPS: Healthcare Cyber – Physical System Assisted by Cloud and Big Data, IEEE Systems Journal, 11, 88 – 95, 2017.

[69] Ahmadi H. , Abdelzaher T. F. , and Gupta I. Congestion control for spatio – temporal data in cyber – physical systems, in Proc. 2010 ICCPS, Stockholm, Sweden, 2010, pp. 89 – 98.

[70] Xue C. J. , Xing G. , Yuan Z. , Shao Z. , and Sha E. H. Joint sleep scheduling and mode assignment in wireless cyber – physical systems, in ICDCS Workshops, Montreal, Quebec, Canada, 2009, pp. 1 – 6.

[71] Scheuermann C. A Metamodel for Cyber – Physical Systems, Technischen Universitä̈t München, DSc dissertation, Munich, Germany, 2017.

[72] Ma L., Yuan T., Xia F., Xu M., Yao J. P., and Shao M. A high-confidence cyber-physical alarm system: design and implementation, in Proc. Intl. Conf. on Green Comp and Comm and Intl. Conf. on Cyber, Physical and Social Computing, 2010, pp. 516-520.

[73] Hu F., Lu Y., Vasilakos A. V., et al. Robust Cyber-Physical Systems: Concept, models, and implementation, Future Generation Computer Systems, 56, 449-475, 2016.

[74] Li W., Meng W., Sakurai K., and Kwok L. F. Towards False Alarm Reduction Using Fuzzy If-Then Rules for Medical Cyber Physical Systems. IEEE Access, 6, 6530-6539, 2018.

[75] Laghari A. A., He H., Karim S., Shah H. A., and Karn N. K. Quality of Experience Assessment of Video Quality in Social Clouds, Wireless Communications and Mobile Computing, 2017, 2017.

[76] Liu Y., Cheng S., and Hsueh Y. eNB Selection for Machine Type Communications Using Reinforcement Learning Based Markov Decision Process, IEEE Transactions on Vehicular Technology, 66, 11330-11338, 2017.

[77] http://www.interdrone.com/news/researchers-to-test-a-drone-traffic-management-system

[78] https://www.faa.gov/uas/research/utm/

[79] Alnasser A., Sun H., Jiang, J. Cyber Security Challenges and Solutions for V2X Communications: A Survey, Computer Networks, 151, 52-67, 2019.

[80] Healthcare Global, 2013, Remote health monitoring. [Online]. Available: http://www.healthcareglobal.com/sectors/medical-devices-products/3g-remote-health-monitoring-facilitated-sprint-and-bl-healthcare

[81] Le Callet P., Möller S., and Perkis A. Qualinet white paper on definitions of quality of experience. European Network on Quality of Experience in Multimedia Systems and Services (COST Action IC 1003) 3, 2012.

[82] Laghari A. A., Sadhayo I. H., and Channa M. I. Enhanced Autonomic Networking Management Architecture (ENAMA), Engineering, Science and Technology, 9, 2015, 1-5.

[83] Dillon T. S., Zhuge H., Wu C., Singh J., and Chang E. Web-of-things Framework for Cyber-Physical Systems, Concurrency and Computation: Practice and Experience, 2010. doi:10.1002/cpe.1629.

[84] Singh J. Multicast QoS routing using collaborative path exploration, in Proc. 24th IEEE Int'l Conf. Advanced Information Networking and Applications (AINA), Australia, 2010, pp. 120-125.

[85] Laghari A. A., Rehman K. U., and Connelly K. Toward Total Quality of Experience: A QoE Model in a Communication Ecosystem, IEEE Communications Magazine, 50(4), 58-65, 2012.

[86] Rehman Z. U., Altaf S., and Iqbal S. Survey of authentication schemes for health monitoring: a subset of cyber physical system, In Proc. 16th Int'l Bhurban Conf. Applied Sc. and Techn. (IBCAST), Islamabad, Pakistan, 2019, pp. 653-660.

[87] Coelho A. M., Estrela V. V., Carmo F. P., and Fernandes S. R. Error conceal-ment by means of motion refinement and regularized Bregman divergence, in Proc. 13th Int'l Conf. Intelligent

Data Eng. and Automated Learning, Natal, RN, Brazil, 2012.

[88] Fernandes S. R., Estrela V. V., Magalhaes H. A., and Saotome O. On improving sub-pixel accuracy by means of B-Spline, in Proc. 2014 IEEE Int'l Conf. Imaging Systems and Techniques (IST 2014), Santorini, Greece, pp. 68-72, ISBN: 9781479952199.

[89] Hemanth J., and Estrela V. V. Deep learning for image processing applica-tions. Adv. Par. Comp. IOS Press, Netherlands. ISBN978-1-61499-821-1 (print) 978-1-61499-822-8 (online) 2017.

[90] de Jesus M. A., and Estrela V. V. Optical Flow Estimation Using Total Least Squares Variants, Oriental Journal of Computer Science and Technology, 10, 563-579, 2017. doi: 10.13005/ojcst/10.03.03.

[91] Coelho A. M., and Estrela V. V. State-of-the-Art Motion Estimation in the Context of 3D TV, Ed.: Reuben A. Farrugia, Carl J. Debono. (Ed.). Multimedia Networking and Coding. 1ed. Malta: IGI Global, Hershey, PA, USA, 2013, pp. 148-173, ISBN: 1466626607.

[92] Ahmed E., and Gharavi H. Secure Cooperative Communication, IEEE Transactions on Intelligent Transportation Systems, 19(3), 996-1014, 2018. doi:10.1109/TITS.2018.2795381.

[93] Javed M. A., and Hamida E. B. On the Interrelation of Security, QoS, and Safety in Cooperative ITS, IEEE Transactions on Intelligent Transportation Systems, 18(7), 1943-1957, 2017.

[94] Yaacoub E., Filali F., and Abu-Dayya A. QoE Enhancement of SVC Video Streaming Over Vehicular Networks Using Cooperative LTE/802.11p Communications, IEEE Journal of Selected Topics in Signal Processing, 9(1), 37-49, 2015.

[95] Jailton J., Carvalho T., Araújo J., and Frances R. Relay Positioning Strategy for Traffic Data Collection of Multiple UAVs Using Hybrid Optimization Systems: A FANET-Based Case Study, Wireless Communications and Mobile Computing, 2017. doi: 10.1155/2017/2865482.

[96] Sahingoz, O. K. Networking Models in Flying Ad-hoc Networks (FANETs): Concepts and Challenges, Journal of Intelligent & Robotic Systems, 74(1-2), 513-527, 2014.

[97] Marconato E. A., Maxa J. A., Pigatto D. F., Pinto A. S. R., Larrieu N., and Branco K. R. L. J. C. IEEE 802.11n vs. IEEE 802.15.4: a study on com-munication QoS to provide safe FANETs, in Proc. 46th IEEE/IFIP Int'l Conf. on Dependable Systems and Networks, DSN-W, Toulouse, France, 2016, pp. 184-191, 2016.

[98] Rohde S., Putzke M., and Wietfeld C., Ad hoc Self-healing of OFDMA Networks Using UAV-based Relays, Ad Hoc Networks, 11(7), 1893-1906, 2013.

[99] Tong W., Hussain A., Bo W. X., and Maharjan S. Artificial Intelligence for Vehicle-to-Everything: A Survey, IEEE Access, 7, 10823-10843, 2019.

[100] de Jesus M. A., Estrela V. V., Saotome O., and Stutz D. Super-resolution via particle swarm optimization variants. In: Hemanth J., and Balas V. (Eds) Biologically Rationalized Computing Techniques For Image Processing Applications. Lecture Notes in Computational Vision and Biomechanics, vol 25. Springer, Cham, Zurich, Switzerland, 2018. DOI: 10.1007/978-3-319-61316-1_14

[101] Razmjooy N, and Estrela V. V. Applications of Image Processing and Soft Computing Systems in Agriculture, IGI Global, Hershey, PA, USA, 2019. DOI: 10.4018/978-1-5225-8027-0.

[102] Razmjooy N., Estrela V. V., and Loschi H. J. A Study on Metaheuristic – Based Neural Networks for Image Segmentation Purposes, Data Science Theory, Analysis and Applications, Taylor and Francis, Abingdon, UK, 2019.

[103] Agrawal J., Singhal A., and Yadav R. N. Multipath routing in mobile Ad – hoc network using meta – heuristic approach. In Proc 2017 International Conference on Advances in Computing, Communications and Informatics (ICACCI), Karnataka, India, 1399 – 140, 2017.

[104] Razmjooy N., Ramezani M., and Estrela V. V. A solution for Dubins path problem with uncertainties using world cup optimization and Chebyshev polynomials, in: Y. Iano *et al.* (Eds), Proc. 4th Brazilian Techn. Symp. (BTSym'18). BTSym 2018. Smart Innovation, Systems and Technologies, Campinas, SP, Brazil, vol 140. Springer, Cham, Zurich, Switzerland, 2019. DOI: 10.1007/978 – 3 – 030 – 16053 – 1_5.

[105] Gaggero M., Paola D. D., Petitti A., and Caviglione L. When Time Matters: Predictive Mission Planning in Cyber – Physical Scenarios, IEEE Access, 7, 11246 – 11257, 2019.

[106] Atat R., Liu L., Wu J., Li G., Ye C., and Yang Y. Big Data Meet Cyber – Physical Systems: A Panoramic Survey, IEEE Access, 6, 73603 – 73636, 2018.

[107] Sztipanovits J. Model integration and cyber physical systems: A semantics perspective, in Proc. Intl. Conf. on Formal methods, 2011.

[108] Singh J., Veeraraghavan P., and Singh S. QoS Multicast Routing Using Explore Best Path, Computer Communications, 29(15), 2881 – 2894, 2006.

第 11 章　总结与展望

目前,对无人机的认识不仅促进了军用,也促进了民用。飞行器的要求是保证更高的安全性,应当与有人驾驶的飞机相媲美。探测车辆路径中的障碍物并确定其是否构成威胁的过程,以及避免这些问题的措施,称为"看见并避免"或"感知并避免"。其他类型的决策任务可以通过计算机视觉和传感器集成来完成,因为它们有极大潜力来提高无人机的性能。从宏观上看,无人机是一种网络-物理系统(CPS),尽管存在着精度、可靠通信、分布式处理能力和数据管理等严格的设计约束,但仍可以从各种类型的传感框架中获益。

本书关注的是 UAV-CPS 领域中仍然存在的几个问题。因此本书讨论了一些趋势和需求,以引起读者的批评,并提供进一步的思考空间。

UAV-CPS 的一些显著优点是,它们能够在完成人类琐事的同时承受人类的职责和目标。同时,它们也独立地做一些决定和执行一些动作。因此,人和机器必须协作。尽管这些功能提供了可观的回报,但要完全掌握辅助人机交互的合适方法,仍然需要付出巨大的努力。

使用低成本的开源组件以及多个传感器和执行器在工作量和成本方面都是一个相当大的挑战。因此,在 UAV-CPS 中使用开源软件(OSS)和开源硬件(OSH)是很有必要的。只要有可能,OSH 和 OSS 应该独立于硬件框架以及操作系统的类型进行设计。

正交频分复用(OFDM)UAV-CPS 的性能可以通过添加信道编码(纠错码)来提高,以识别并纠正在数据传输过程中发生的错误。与几种微型飞行器(MAV)应用中的常规无线信道相比,超宽带(UWB)信道的巨大带宽可以产生新的效果。基于 IEEE 802.15.4a 的 UWB 技术在需要使用各种传感器进行高精度定位以实现稳定和导航时有多种用途。然而,绝对的室内定位依然对无人机的设计与发展构成挑战。

在多变的非结构化环境下驾驶无人机具有挑战性。为了支持更好的算法开发,建议将给定的巴西环境中针对低空无人机飞行的多用途数据集,作为定位和其他航空电子任务的基准,以评估计算机视觉程序的稳健性和通用性,并提供有无地标的深度估计基线。这一发展阶段有助于推进未来与遥感(RS)模块进行集成,从而为分析带来更多光谱信息。

UAV-CPS 涉及大量的网络知识,尤其是飞行自组织网络的知识。UAV-CPS 进行的高维多媒体数据流量呈指数级增长,根据这一事实本章提出了若干问题,并指出了未来的研究方向。

纹理是识别图像中物体或感兴趣区域(ROI)的一个重要特征,它通常用于卫星图像的分类,例如评估生物量。无人机图像拥有超高的空间分辨率,这表明纹理也是最重要的知识来源。尽管如此,无人机图像中的纹理很少用于监视。此外,合并地面高光谱数据可以补偿无人机传感器的有限频带,并提高分析的估计精度。因此,本章的目标是:① 探索基于无人机的多光谱图像;② 通过高光谱信息提高几种类型的估计精度。

配备摄像头的无人机用于直观观察生态平衡、水体、建筑物、桥梁、森林保护区和其他类型基础设施安排的建设和运行,这些应用呈指数级增长。这些 UAV-CPS 可以经常检查各种场所,监控正在进行中的工作,生成安全文件/报告,并检查现有结构(主要是难以到达的区域)。无论机载传感器的复杂程度如何,云技术、遥感技术、计算智能技术和通信技术多么先进,在相当长的一段时间内,对超分辨率的需求都会越来越高。在医疗保健、天文学和救灾等昂贵而又麻烦的图像采集中,这样的需求都会持续增长。QoS 和 QoE(包括其他定性性能指标)都将发挥关键作用,以推动 UAV-CPS 在各阶段的进一步改进。

本书旨在为当前和未来的 UAV-CPS 应用提供参考。它展示了无人机成像能力和传感器集成的基本方面,正在进行的研究工作、成果和面临的挑战。

简而言之,本书详述了计算机视觉和图像处理框架以及无人机设计中传感器的挑战、角色、技术问题和应用。本书各章节将帮助读者专注于与计算机视觉和传感相关的最关键因素。

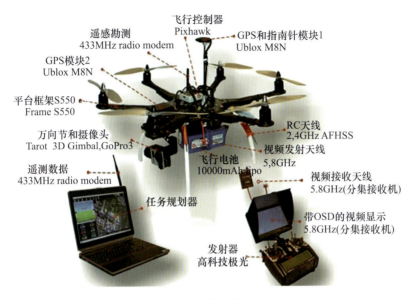

图 3.2 OSH 和基于软件的 UAS

图 3.6 WebODM 用户界面:点云模型

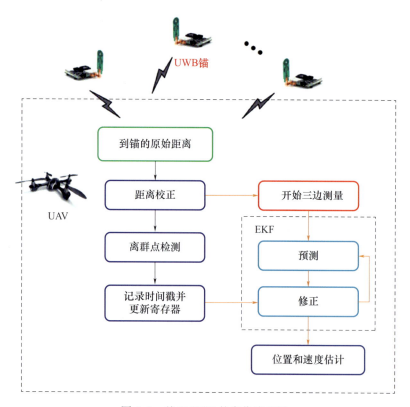

图 4.1 基于 UWB 的定位流程图

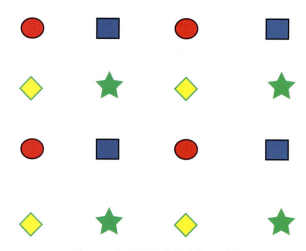

图 4.3 空间域图像的多相子采样

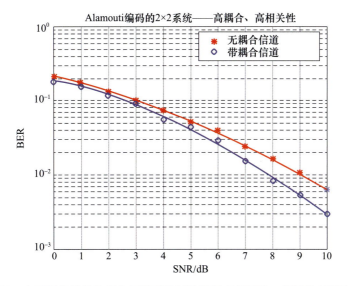

图4.10 在不同的相关和耦合情况下绘制的误码率(BER)–信噪比(SNR)曲线

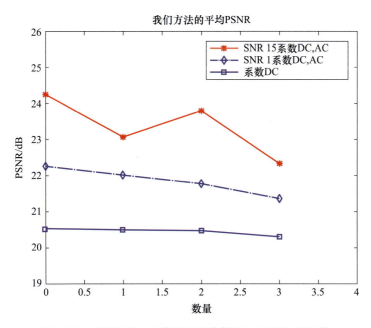

图4.12 不同版本丢失信号的峰值信噪比(PSNR)平均值

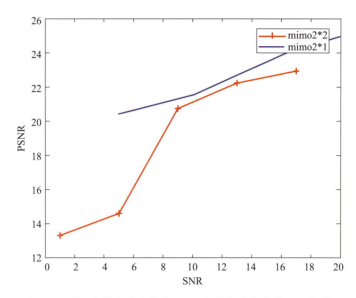

图 4.14 利用离散余弦变换的 MDC 方法的峰值信噪比 - 信噪比

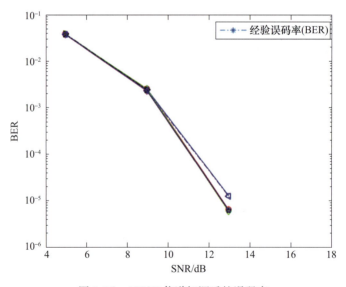

图 4.15 AWGN 信道解调后的误码率

彩插

图 5.1 路径（绿色针脚表示交叉点，按升序编号，以指示路线的方向）

(a) 1:00 p.m.　　　　(b) 4:00 p.m.

(c) 5:00 p.m.　　　　(d) 6:00 p.m.

图 5.2 光照变化示例

图 5.3 黏土砖面具

图 5.4 用差分 GPS 测量土壤中的支撑点

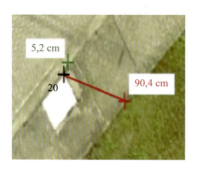

图 5.9 用差分校正 GPS 采集的地面坐标(黑色标记)、用土壤中的支撑点安装的镶嵌图(绿色标记)和仅用无人机 GPS 数据安装的镶嵌图(红色标记)之间的误差

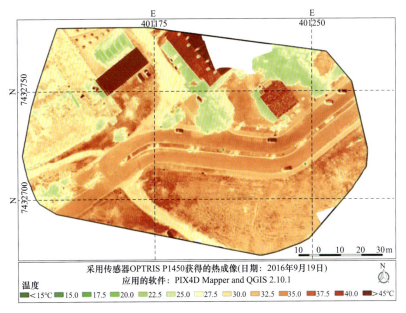

图 5.11 热光谱镶嵌

(a) 环形交叉路口地标　　　(b) 屋顶的识别　　　(c) 对屋顶的识别

图 5.16　GPS 自动定位位置比较

（地标识别估计，与无人机在帧中的实际位置相关。红点是地标识别系统估计的位置；
绿点是 GPS 估计值；黄点就是真实的位置）

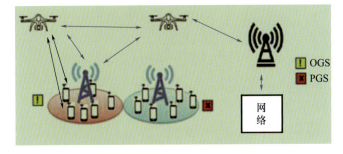

图 6.1　无人机作为网络网关

（OGS:过载的 GS；PGS:有问题的 GS）

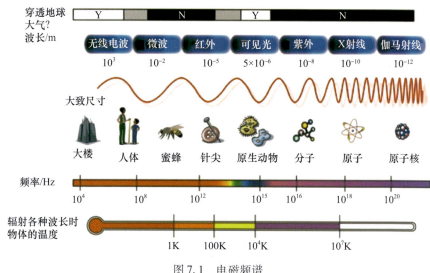

图 7.1 电磁频谱

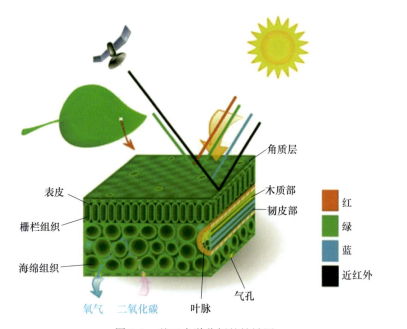

图 7.2 基于光谱分析的植被图

彩插

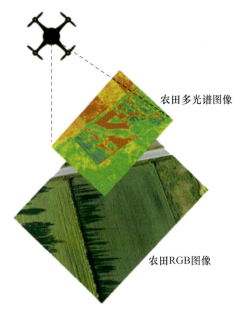

图 7.5　多光谱成像

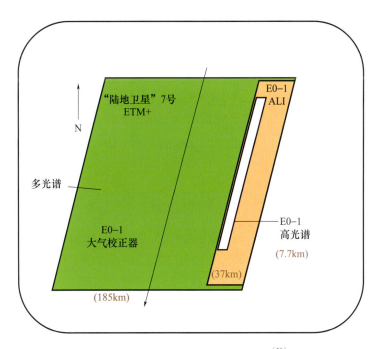

图 7.6　空间分辨率：MSI 比对 HSI[30]

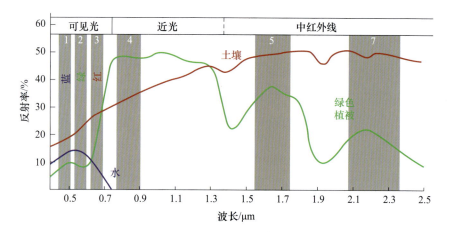

图7.7 光谱反射率图[41]

图10.9 UTM(1英尺=30.48cm)